Effective Teaching

Effective Teaching:
CRITICAL SKILLS

BY

Stanley T. Dubelle, Ph.D.

LANCASTER · BASEL

Published in the Western Hemisphere by
Technomic Publishing Company, Inc.
851 New Holland Avenue
Box 3535
Lancaster, Pennsylvania 17604 U.S.A.

Distributed in the Rest of the World by
Technomic Publishing AG

While many organizations and persons mentioned in this book are real, certain organizations and persons, particularly those in examples, are fictitious, and any resemblance to actual organizations or to actual persons, living or dead, is purely coincidental.

Printed in the United States of America
10 9 8 7 6 5 4 3 2 1

Main entry under title:
Effective Teaching: Critical Skills

A Technomic Publishing Company book
Bibliography: p. 139

Library of Congress Card No. 86-51053
ISBN No. 87762-487-9

To Jane Dixon Dubelle,
my wife for all seasons.

Table of Contents

Preface

A number of years ago a creature from out of this world came to visit us on our American movie theater screens. It was The Extraterrestrial, better known as E.T.

While E.T. may not have been from this world, this "E.T." book, *Effective Teaching,* is definitely from this world. The book represents a sincere attempt to communicate a down-to-earth message on what it takes to make teaching effective. And teaching is effective only when it is borne out in student learning.

The teaching skills and ideas presented in *Effective Teaching* are a translation of the work of the educational researcher, whose written reports are oftentimes hard to understand. Also, the skills and ideas in *E.T.* are complementary with proven teaching theory and principles of human psychology.

Finally, the skills and ideas have been tested and proven in the classroom setting by hundreds of teachers, including the author . . . as many a teacher remarked, "This stuff really works!" It certainly should, for the ideas have been borrowed freely from successful practitioners, researchers and theorists.

The educational institution is under heavy scrutiny today, but most of what is being investigated is the curriculum. Curriculum is the WHAT of the student's education. Strangely enough, the blue ribbon panels and the Presidential Commissions are forgetting to look at our delivery system, the HOW of education, the act of teaching. The HOW has always been a force to reckon with in American education. That force has to be put in capable, professional hands—the hands of you, the teacher.

This book on the HOW has been divided into three Parts. Part 1 deals with the act of teaching that is introductory to the main body of teaching. As such it deals with the preparation of appropriate objectives, understanding of sound motivational practices and effective set induction activities.

Part 2 deals with the "middle" of the teaching act which is to say that it deals with what the effective teacher must do once the lesson is underway and before it is ended. Five teaching skills are identified and analyzed in this part: reinforcement cueing, encouragement, enthusiasm, stimulus variation and questioning.

Part 3 focuses on closure as a culminating teaching skill—culminating in the sense of honing in on major parts of the lesson as well as the "finale" of the lesson.

As indicated earlier, the delivery system for the curriculum is the HOW of education or the act of teaching. It is the teacher who "delivers" the curriculum to the student. Teaching as a delivery system is a force, a force the effective teacher must reckon with.

May the *E.T.* force be with you.

Stanley T. Dubelle, Ph.D.

Acknowledgment

Carol M. Hoffman read the manuscript for this book and I would like to acknowledge a debt of gratitude to her for doing so. Her comments, suggestions for improvement, and rock-bottom friendship are invaluable assets for the trials and tribulations that were involved with the writing of this book. Everyone should have a C. H. in their life.

PART I

Beginning the Lesson

Chapter 1

FOCUS ON EFFECTIVE TEACHING

You have chosen the most important profession in the world—TEACHING! It is the queen of all occupations. It is the mother. From teaching, all other occupations and all other professions are born. No matter how long you commit yourself to teaching, you must always have that sense of special pride in being a part of something that influences all other things. Henry Adams said it best: "A teacher affects eternity. He can never tell where his influence stops." Such is the special place of the professional educator, you, the teacher.

Chapter Objective

Understand how the development and use of one's teaching skills are vital to effective teaching.

The first and most important work of a teacher is to get students to want to learn whatever it is that's about to be taught. First graders have to want to learn to read; tenth graders have to want to learn biology; twelfth graders have to want to learn calculus. Sounds simple but it really isn't. Human beings are very complex. Each one has a variation on the theme of what an individual wants, or needs for that matter. Yet in beginning a lesson, the teacher must plan an introduction that reaches the wants or needs of all the students in the class.

There are several things that students have in common, things that we can use to get them to want to learn. One of them is the desire to be competent in their world, competent in whatever they are expected to do.

Think about yourself. If you are an experienced teacher or a person about to go into teaching, there's one thing that's true about you. You want to be successful with your students. You do not want to flop or be ineffective. Wanting to fail as a teacher is masochistic, and there just aren't many masochistic teachers around.

Because you are a professional educator or you are about to become a professional educator, you have a striving to be competent as a teacher. The fact that you are reading a book on effective teaching practices is evidence of that point. If you had a teacher, or you know of a teacher, who really did not care about competence, then you experienced a rarity. Such individuals are in a distinct minority and certainly are not true members of the *profession* of education.

Just as you want to be effective as a teacher, your students want to be effective in mathematics, social studies, physical education, communication, art, music, spelling, handwriting, typing, woodworking, sewing, science, and whatever else wasn't mentioned. Students may protest to the contrary with, "Who needs this stuff anyway!" or "Yuk! I hate spelling!" Students tend to enjoy protesting and griping. But make no mistake about it, they do want to master the subject or at least they *did* at one time before the aftereffects of frustration set in. In any case, you can tap into that desire, that wanting to be competent. That's where this chapter taps in to you—you wanting to be more effective.

Since you are a professional, or you are on your way to becoming one, you are probably striving to be more effective. Our profession needs "strivers" like you.

In Chapter 2, some specifics of motivational teaching practices will be identified. For now, I hope that what you have read has prompted you to *want* to read on.

As stated earlier, the first work of the teacher is getting students to want to learn. That's where the planning and preparation for effective teaching begins.

Effective teaching, most simply put, has an attractive beginning. It also has a supportive middle and a productive ending. Of course that's oversimplifying the act of teaching, but we need to begin somewhere. Reduced to its absolute essence, the organization of the act of teaching is just that . . . a beginning, a middle and an ending.

The teaching skills we consider in this book will be identified for each time-segment of the teaching act. However, they can be used at any time the teacher finds them appropriate. As you would expect, there is overlapping in the skills; they are not mutually exclusive. The time when the skills are to be used and the manner in which one skill becomes a part of another will become clearer as we progress through the book.

Looking at teaching from an overview perspective, several points need to be made. Undoubtedly, the most critical point is that teaching at its best is a matter of one person teaching others. That *one* person is a human being called teacher. The others are human beings called students. Teaching is a human encounter. There isn't another human encounter quite like it.

In more than 30 years of experience in the field, I have been observing a discomforting factor in the work of teachers. It is the practice of teachers making and implementing plans that are first, last and always based on subject matter. Many teachers have traditionally wanted to get the "material" covered or the book finished or to be at a certain place at a certain time. Such activity is symptomatic of an indulgence in subject matter.

Subject matter is certainly vital to any school program. From it we draw the knowledge and the skills that must be imparted to our students. Students must master the knowledge and the skills in order to be productive, contributing members of adult society. But the orientation of teaching should not have subject matter as its first order of business. Its first order of business must be the learner.

We like to say that "we teach children" or "we teach junior high students." Truth is, we teach reading, writing, foreign language, and mathematics. We teach subject matter as a first priority. The students we teach tend to be a second or even a third priority. Sometimes the push to get to a certain place in the course or subject takes precedence over plans that ought to be made for the learners' needs.

The developmental levels of learners, their learning styles, and their needs/interests/abilities must all be considered. The teaching skills that are identified, explained, and exemplified in this book are based on a learner-orientation first, a subject-matter orientation second.

Another point that needs to be made is that teaching is too often done on a fly-by-the-seat-of-the-pants basis. It is often done spontaneously. Some teachers just get the lesson started and take it from there, or they follow the plans provided at the end of the textbook's lesson. In all of these cases there is an absence of teaching strategy—a strategy that would make it more likely that learning will occur.

The teaching skills in this book are organized and based on the fact that when used they will make learning more likely to occur than if no teaching strategies were used. We could even raise the question as to whether unplanned teaching is really teaching at all. If students do learn, it may be a case of students learning in spite of the teacher. How would you like it if your students were said to have learned "in spite of you"? It would hurt, wouldn't it?

Still another point to be made is that there are too many errors made in teaching. There are errors of omission as well as errors of commission. Examples: The teacher never informs the students what the lesson objectives are (omission); the teacher teaches the entire lesson sitting behind the desk (commission). Such errors can be avoided by teachers who know and have a mastery of the effective skills to use in certain situations.

The skills in this book represent neither the final nor the total word on what constitutes error-free teaching. However, they are a valid beginning for teachers who would like to be more effective. They are skills that are universal for all levels of teaching, from kindergarten through twelfth grade (and even beyond).

One critical error in teaching is often made at the beginning and the ending of a lesson. Some lessons actually have no *planned* beginning and no *planned* ending. The teacher simply begins somewhere, usually where the class left off yesterday. It's a "plunge right in" situation. Sometimes it's analogous to plunging the unsuspecting learner into a cold shower. Such plunges are seldom invigorating for the students.

In the secondary school it is often a case of the bell ringing and ending the lesson. In the elementary school it is often a case of time running out and the "Bluebirds" reading group lining up for their turn with the teacher, who is working in the reading corner. The ending of a lesson by anything other than planned teacher action is highly inefficient and ineffective.

Beginning and ending lessons efficiently and effectively . . . what can be more logical and obvious than that? In the following chapters we will deal with effective skills to use in beginnings and endings as well as effective skills to use in between.

Speaking of in between teaching skills, it is truly disheartening to observe teachers droning through their thirty- to forty-minute lessons. Worse yet, is observing the "drone-ees," the students who must sit through boring lessons that are infected by flat teaching. The saddest part of the indictment of monotonous teachers is that their teaching, their lessons do not have to come across in such a dull way. With a little commitment by the teacher to work on certain skills, dullness can give way to dynamism. Developing effective teaching skills is more energizing than taking megavitamins or B_{12} shots.

One final point. As mentioned in the Preface, this book is written to form a bridge between the educational researcher and the classroom teacher. Considerable research on effective teaching has been done, much of it sound. It is in the reporting of the results that problems appear. Whether the reports appear in periodicals or books, the language is typically that of the researcher, replete with statistical jargon and terminology. The jargon and terminology are unfamiliar to most teachers.

The criticism is not of the researcher, but of the fact that little is written to link what the researcher says and what the teacher reads and does. The teaching skills in this book are research based. While there will be little reference made to research studies, you can be confident that the skills are verified by research studies as valid effective teaching skills.

The hope is that you will sense the logic and the common sense appeal for the use of the skills. Having sensed that, the next hope is that you will commit yourself to their use.

To the educational supervisors who may be reading this book, here's a suggestion. Supervisors must know and be able to do many things. You can do nothing more important than to master the teaching skills that make for effective teaching.

The best supervisors are not masters of teachers; they are master teachers. The only way to be a master teacher is to achieve personal competence in using teaching skills. Knowing the skills, understanding the skills, and being able to critique teachers using criteria related to the skills are all fine. However, in the finest sense of supervision, a supervisor must be able to demonstrate teaching competence.

The supervisor does not have to be better than the teachers he or she supervises in the use of the skills. That's an unrealistic expectation. Not even the teacher being supervised would expect that.

The supervisor must be able to demonstrate the teaching skill in question for several reasons. First, supervisors must have credibility if teachers are to respond to their lead. Second, teachers need competence models to observe and emulate. Third, the teacher needs valid criticisms from someone in the field who can combine the skill of teaching with the subject matter at hand and offer constructive analyses (it is unrealistic to expect school principals to be knowledgeable in the various subject fields). Fourth, teachers are tired of getting advice from supervisors (and administrators) who talk a good game but never put their teaching to the test.

Teachers appreciate supervisors who are willing to demonstrate what they are talking about—even if they, too, do not do it right the first time out. Teachers need supervisors who care enough about them to get into their "teaching shoes" every now and then. Teachers want supervisors who care enough to feel and experience the challenge of today's students. Caring enough about the teacher to demonstrate the skill is what counts. A commitment to mutual respect in an air of cooperative action is the desired relationship between teacher and supervisor.

The next chapter is devoted to getting students to a place where they want to learn. As stated earlier, getting students to want to learn is a matter for motivational skills inherent in teaching.

Let's hope there's a better way to get students motivated without using carrots and clubs, rewards and punishments, grades, smiley stickers, gold stars, etc. There is! See if you can guess what the most powerful motivator inside the learner is. . . .

Chapter 2

t_cCIR (MOTIVATION)

Curiosity, that's what! Let's hope you were curious about the answer to the last line of Chapter 1. If you forgot the question, go back and look at that entire last paragraph. If you don't, the first two sentences of this chapter won't make sense. Making sense *together* is a must for us!

Speaking of curiosity, are you curious about the name of this chapter? Look again at those letters. Strange combination. If you are very liberal in your pronunciation attempts, you can almost force the T-E-A-C-H-E-R sound out of it. t_cCIR is not an acronym, but the remembering of those five symbols, t, $_c$, C, I, R can prove helpful to you as a teacher. Mnemonics is a great help in remembering. Remembering is a great help in understanding.

Understanding the ideas behind the five symbols will provide you with insights into what truly motivates students. That's where the help comes in. The ideas or concepts, represented by the five symbols, form the basis for sound intrinsic motivational theory.

Chapter Objective

Understand the concepts and principles of intrinsic motivational theory.

Before we get into the concepts of motivation, we need to arrive at a working definition of intrinsic motivation. We can probably do that best by contrasting it with extrinsic motivation, which is probably familiar to everyone who ever spent time in a formal education classroom.

A working definition of intrinsic motivation would go something like this. Intrinsic motivation is that force which originates *inside* the student and, *by the student's decision,* the force acts on the student attracting him or her to the learning task.

Extrinsic motivation is that force which originates *outside* the student and acts on the student in such a way that he or she behaves in a way that is acceptable to the person controlling the force. Teachers activate the force to get students to do what they want them to do. Extrinsic motivation is forceful and coercive—even when it is used gently and lovingly! Let's take an open-minded look at that last idea.

Suppose one of your students responds correctly to something you taught her. You are a devout follower of what you learned in Psych. 101 and Ed. Psych. 201. You know that whenever the student responds in a desirable way, your job is to make sure that she experiences some pleasurable event immediately after her response. You reward her with a pleasurable "event" like a gold star, an extrinsic motivator. Having accepted the reward, your student will be more likely to respond again in the approved way. Furthermore, the pleasurable aspect of your reward to her will somehow "stamp in" this correct response and she will have learned. So goes the stimulus-response theory as applied to the classroom.

If you reward students for every response they make, you will soon extinguish the very behavior you hope will continue! Even the behaviorists agree with that idea. Besides, you couldn't reinforce every correct response of every student. Even those of you possessed by a missionary zeal to do so couldn't do it. There are too many instances of positive behavior for a human teacher to reward. Come to think of it, even programmed devices like teaching machines, programmed textbooks or computers can't do it either. That's one of the problems—rewarding the myriad of positive student behaviors that occur in the classroom.

Another problem, and a very common one, occurs whenever the teacher rewards positive student responses and the student likes the rewards. Of course, the teacher expects the student to like rewards—but there's a catch to that. He likes it so much that he works even harder to get another. And then another. Well, that's great isn't it? That's what rewards are supposed to do—motivate students to work harder. Working harder will produce more learning, or so the thinking goes.

At first thought, that argument is attractive. But after awhile the teacher finds that the student is working to get *the reward and not the learning.* There is a BIG difference in the purpose of the student's behavior.

What we want is for our students to strive to learn what we are teaching so they attain a mastery over the task at hand. Attaining mastery gives students a natural sense of achievement. Confidence sets in. An urge to repeat the successful task or to go on to another slightly more challenging task also sets in. This is the intrinsic motivation that we want to overtake our students.

When students strive to attain the gold stars, the smiley stickers, the M&M's, the tokens, the grades, the teacher's approval, and all the other extrinsic motivators, they misdirect their efforts. Students will work like a veritable donkey to get the carrot, so to speak. In doing so, they may accomplish some of the learning tasks along the way. However, there is a tremendous waste connected to students' misdirected efforts.

Many students will end up working harder to get the reward than they would have had to work in order to learn the task at hand. The learning becomes *incidental* to the striving for the reward. Can you just see twenty-five elementary children being told that if they finish their last twelve worksheet problems before 1:00 P.M. they can have ten extra minutes of recess time! They'll blaze through that assignment in no time flat just to get that ten minute "carrot." The quality of work won't be a concern for them. Then there's the side effect of the student who is not adept at the particular worksheet task. He gets frustrated when he can't work as quickly as the rest. You know what happens with him. When we reward speedy completion of tasks, the quality of work plunges and the slower-moving thinker becomes frustrated and gives up.

To illustrate the point of the corrupting effects of incentives let's turn to an interesting experiment. Zoologist, Desmond Morris, once did an experiment that exposed an ape to an extrinsic motivator—peanuts. He had gotten the ape to draw and paint. It was doing some very nice work. Then Morris began rewarding the ape with peanuts. "Soon it was doing any old scrawl to get the peanuts," Morris said. "I had introduced commercialism into the ape's world, and ruined him as an artist." Apes and students are not the same. However, they do respond similarly when it comes to incentives for work the teacher wants done.

Staying with extrinsic motivation for one more point, let's look at the flip side of carrots—clubs. While the carrot is the reward teachers give as an incentive for the student to learn, the club is the punishment teachers use on the student who acts like he doesn't want to learn.

If the student misbehaves, he takes himself and others away from the learning task at hand. If he does work on the task at hand and the work is done badly, he is not performing satisfactorily. In either of these cases, punishment is the response of the S-R teacher.

Punishment takes the form of lower grades, scoldings, loss of privileges, things like that. Such things are the province of some teachers and some schools. That matter is not for review in this book. Here it is merely noted that the practices mentioned are extrinsic to the learner. An additional point is that these practices are not nearly so effective as the intrinsic motivators that are also available to the teacher.

Rewards and punishments, carrots and clubs, call them what you will, are the ill-conceived extrinsic motivators currently being used by teachers—too many teachers. They are not at all effective for engendering self-reliant, independent-working, highly-motivated students.

Today's teachers are greatly concerned that their students are apathetic. Ineffective motivation is part of the reason for apathy. We must do something about that. This chapter includes a valid alternative.

At this point, we need to explode some myths about motivation, because certain pseudo-motivational procedures still have a hold on many teachers. The following explosions occurred in the fields of research.

MYTH #1: You can motivate people with a carrot—or with a stick.

FINDING: People are motivated from within themselves. We can only elicit motivated behavior by creating conditions that increase the probability of its occurrence.

MYTH #2: We can provide a person with motivation.

FINDING: Motivation already exists within a person. A good teacher changes factors in the environment, giving motivation a chance to express itself.

MYTH #3: Incentives and bonuses motivate people.

FINDING: Such things can only reinforce motivation. If the motivation isn't there to begin with, the prospect of a reward won't necessarily bring it into existence.

MYTH #4: Eliminating stress and conflict improves productivity.

FINDING: Research shows that it reduces productivity. A moderate degree of tension, that is stress and conflict, enhances productivity.

MYTH #5: People are motivated by praise.

FINDING: Praise is useful only when it reinforces the real satisfaction that comes from accomplishment. Some people are embarrassed by praise. To others it's a form of condescension.

In summary then, the research findings on *extrinsic* motivation provide some clear directions for teachers: (1) the teacher must establish a classroom environment that stimulates and supports the motivation that is innate to the student; (2) at best, incentives and praise may reinforce the intrinsic motivation of the student; (3) the teacher must induce a moderate degree of tension in the classroom environment, if it is not already there. (Note: if the tension is excessive, then the teacher's job is to bring it back to a moderate degree.)

Before getting into the intrinsic motivational concepts of t_cCIR, don't forget our definition: intrinsic motivation is that force which originates *inside* the student and, *by the student's decision,* acts on the student, attracting him or her to the learning task.

t = Tension

Now we're ready for the intrinsic motivation concepts identified by t_cCIR. First the "t."

The "t" is for tension, the tension that the teacher must be sure is present once the lesson gets underway. The term we'll use is t-state.* Before we get into the "how" of arousing t-states in learners, please realize that this concept of tension does not include anxiety, worry, fear, or other debilitating emotions. The effective teacher avoids generating those emotions. We use small letter *t*, in-

*t-state is a term credited to Max Werthemier, an eminent Gestalt psychologist.

dicating the desired, moderate tension as contrasted with capital *T*, which signifies excessive tension.

The t-state can appear in the form of many different emotions. It is the teacher's work to evoke those emotions in students at key times in the lesson. Some of the emotions are curiosity, surprise, inspiration, puzzlement, uncertainty, joy, excitement, happiness, hope, titillation, amusement, and pride. Teachers who elicit those kinds of emotions from their students have created the conditions in which intrinsic motivation can occur. We now understand that the kind of tension we hope to arouse is not the Big *T*!

Students react to the emotion they are experiencing in such a way that they want to "satisfy" it. For example, if students are feeling inspired over something the teacher has said, read or done, they will *want* to act on their emotion, to do something about what they are feeling. In that sense they are acting to satisfy their feelings of inspiration.

In another example, the teacher may have done something in the way of an experiment that left the students uncertain or puzzled as to what they saw. They will *want* to satisfy their uncertainty or puzzlement by becoming certain and "un-puzzled."

In a final example, the teacher may have told a funny story to her students. To satisfy the amusement they feel, the students move in the direction of what piqued their laughter or their giggles. They *want* to hear more or learn what it was that made them feel amused. They want more! The dream of every caring teacher is students wanting more.

In each of the examples cited, the emotion experienced was the "fuel" that propelled or moved the students to learn.

In attempting to provide greater insight into intrinsic motivation, we note that the emotion experienced by the student is an emotion *decided* upon by the student himself! The student decides to be happy, to have hope, to be amused, to be curious, etc. That being the case, it is logical that the student has also decided to be attracted to whatever is evoking his emotion. The student has thus decided for himself to be motivated. That's the "inside" of motivation. That's the psychology of it.

A passing word on the teacher and psychology. In the not too distant future, teachers and schools of education that prepare and provide inservice for teachers will come to realize that every teacher must know and understand the psychology of human behavior. It is the only complete way for teachers to be highly effective with their students. The slogan might well be, "Every teacher a psychologist." Two books for teachers written in that vein are *Misbehavin'* and *Misbehavin' II* (see Bibliography section).

Back to emotion. In a very real sense, emotion is the motivational fuel that teachers must learn to plan for and call on, *if* they want their students to be truly motivated.

It Is Far Easier and Far More Effective to Influence a Student to Enter a Motivational t-State Through Affective Rather than Cognitive Means.

The arousal of t-states is primarily an affective consideration. To put it in its simplest form, motivation challenges the teacher to get the students to WANT to learn. John Dewey said the most important work of the teacher is to get the student to feel the *need* to learn. Needs, wants, t-states are each examples of the affective involvement of the student. The *effective* teacher works on the *affective* plane.

c = Curiosity**

In t_cCIR the $_c$ is a subscript of t indicating it is a type of emotion that is included in the tension concept. Curiosity is a primary emotion for teachers to tap. In fact, curiosity may be viewed, practically speaking, as the prototype for intrinsic motivation. It is that effective in attracting the student to the learning task. It is the teacher's greatest motivational ally. Teachers ought to use it—in industrial strength doses!

Students' attention is attracted to something that is unclear, unfinished or uncertain. Attention is sustained until that something is clear, finished or certain. Students like to be puzzled or challenged in a way that they get the feeling they can figure out a solution to the puzzle or challenge. They *want* to satisfy their desire to "find out" the what's, how's, who's, when's, where's, and why's of curiosity-evoking things. Bruner says, "The achievement of clarity or merely the search for it is what satisfies."†

In the next chapter we will study a specific teaching skill that

**Curiosity, plus the following three motivational concepts, are taken from Jerome Bruner's book, *Toward A Theory of Instruction* (New York: W. W. Norton Inc., 1966), pp. 113–138.

†Ibid., p. 114.

relies heavily on the activation of student emotion to generate interest in the learning task at hand. Curiosity will be a primary emotion for the teacher to tap in the use of that skill.

C = Competence

A desire for competence is another fundamental intrinsic motivator. Students want to be competent in dealing with their world. That is particularly true in the student's social world. She must feel a sense of significance to the group, whether the group is family, friends or the third-period class in social studies. The teacher has a key function in helping the student be successful in her striving for significance.

Significance is achieved whenever the student perceives herself as competent at something and worthwhile to the group. There are several ways, positive and negative, through which this perception can take place. The one that concerns us here is the attainment of a level of competence in what is being taught. Saying it plainly, the teacher must make sure the student experiences success in the subject at hand and continues to do so on a consistent basis. Once is not enough! Achieving success prevents failure which ultimately puts distance between the student and the group, as well as the teacher. That usually means trouble, often in the form of discipline problems.

Aside from wanting to be seen as significant to the group, the student simply wants to be successful in what she does. It is an innate striving. However small the increments of success in the subject, the teacher must build on the student's level of competence. Students get interested in what they get good at doing. No one gets interested in something he cannot do.

As the student accomplishes the tasks and goals of the subject at hand, she feels a growing confidence in herself. This is the movement toward competence.

The Competence Motive Is Cyclical. Students Get Interested in What They Get Good at Doing.

There are at least two basic things to consider in presenting subject matter that you'd like the student to master:

(1) There must be meaningful structure to the task. It cannot

be a mishmash of this and that of which the student can make no sense. She has to see the logic, the flow, the emerging Gestalt, the forest as contrasted with the trees.

In that structure of the task there has to be a clear beginning and ending. It is only by seeing that a task has been begun and ended successfully that the student can experience the joy of accomplishment. Of course, learning is a continuous phenomenon and so, too, are the tasks we set for students to learn. But we can divide the tasks into meaningful increments which have beginnings and endings.

The effect we want to see generated in our student is one where the student finishes the task, steps back, looks at it and says, "I done it myself." And don't worry about her grammar! You can work on that task later. (You might even want to say, "Yep, you done it all by yourself, all right!")

(2) The task must require a degree of effort to learn. It cannot be some Mickey Mouse item that has no challenge to it. Worse yet, it can't be something that the student and everyone around her can see is beneath the "regular" student's ability level. That would be analogous to a seventh grade student reading a primary grade book in full view of other seventh graders who are reading books that are written on grade level.

"We learn by doing" is one of education's shibboleths, yet it is definitely appropriate here. Whatever *you* do, make sure the task is a bit challenging, requiring that degree of effort. However, ask students to *do* only what they are developmentally capable of *doing*.

To give a few examples, consider the tests you make for students. Sound test-making practices dictate that you have the student confront some initial questions that can be answered with relative ease, but not without some effort of course. Getting a positive start gets the student "into" the test. And no giving the student one point credit for spelling his name correctly (especially if his name is Otto).

Another example is the preparation of learning centers, which are prevalent in American elementary classrooms. (The secondary equivalent of a learning center would be an independent study project.) When students begin working on their learning center or independent study project they need to experience some early success with their work. Wise teachers build early success into the structure of the task.

A third example is the oral questions a teacher typically asks in class. A sound procedure to use is asking a series of easier questions that will lead up to a difficult one the teacher is planning to ask. The teacher would not always want to do it that way, of course. The teacher may intentionally or unintentionally ask a question that is too difficult to answer right off. Rather than answer her own question, an all too frequent practice, she breaks it down into more manageable increments. Answering the lesser questions successfully, students gain confidence in their ability to answer the previously asked, more difficult one.

Success begets success. Make sure your students get it.

I = Identification

That humans want to identify with other humans almost goes without saying, but we need to say it. We need to think about it because it is another intrinsic motive in the t_cCIR combination.

Why do we look to another person for a model? We do it so we can imitate the trait that the other person has that we'd like to have. We don't yet have the confidence to demonstrate on our own that we have the trait in question. We need a model. We need a good model so we can practice being a better person.

Young people especially look to models. Think of the success the stars of the movies, TV and the world of music have with youth. These stars and their promoters capitalize on the intrinsic motive of identification. Teachers need to understand what those stars and their promoters understand about this motive.

Teachers have the task of providing the right models for their students. This task has two basic components to it, *who* the models are and *what* they model.

As for the who, the most logical and the most important model is the teacher himself. When school districts hire a teacher they not only hire someone who can teach a particular subject to students, but they also hire someone who will serve as a model to students. The teacher is a model to whom the students can look for solid morals, values, direction, etc. Some teachers protest to the contrary saying that they were not hired to teach those things, just a particular subject matter.

The fact is that teachers have *no choice* as to whether or not they will serve as a model for students.

Their only choice is what kind of model. Students will naturally look to the teacher. Like it or not, such is the intrinsic motive of identification.

For purposes of subject matter, the teacher as model must be someone who has achieved mastery in his subject field(s). He must at least know enough about his subject so he can teach it at a level commensurate with the grade or age level he is teaching.

A more desirable teacher-model would be one who goes beyond mere competence. He would be a teacher who is ever a student, always learning about his field, always enthused about his learning. He would show those things to his students. He would especially model the enthusiastic learner. In doing so, students would identify with that enthusiasm and respond positively to it by learning more—much more than just the subject or the minimum requirement. Think of the teachers you had who went beyond mere competence. What influence they had! Teachers can use other faculty as competence models for their students. Models could be seen in faculty resource people like the librarian, the nurse, the guidance counselor, the principal, even the superintendent of schools (little humor there). Such models could be of great help to the regular classroom teachers.

Outside resource people could serve as competence models as well. Teachers must be judicious in their selection, however. Whether we bring resource people into the schools as guest speakers or we send students to them by way of a field trip, the persons with whom the students are to identify need to be "right."

There is one more not-to-be-forgotten competence model, and that is another student. Sometimes we forget that students learn a great deal from one another. That is one of the strengths of education, providing instruction in a social setting. Students *do* learn from one another. Whether by some peer tutoring process or by the day to day interaction among students, the intrinsic motive of identification will work, for better or for worse. The teacher is responsible to make it work for the better.

The Identification Motive Is Absolute. Teachers Have No Choice Whether Their Students Will Identify with Models, They Can Only Choose What Kinds of Models They Will Provide Students.

As for what the models will model, there is the need for students to identify with other students and with adults who have a mastery of the knowledge or skill required. There is the need for identification with models who will provide the right direction in the affective domain of attitudes, values, habits, ideals, morals, etc.

Teachers who provide the right identification models are those who will surely enjoy the fruits of a long successful career. To them the term "burnout" is rightfully relegated to the lexicon of space exploration.

R = Reciprocity

We come to our fifth component, reciprocity. Reciprocity is best defined as the need to interact with others, that is, the need to communicate about items of mutual interest or concern.

Students, like everyone else, function in a social web of interdependence. There is an "ironclad logic" to our social living, as renowned psychologist Alfred Adler put it. Synonyms for reciprocity might be teamwork, social interest, cooperation, esprit de corps, common cause, even misery loves company. There is a bonding of one human being to others of the same kind. It is a fact of human nature on which the teacher can capitalize—especially in the classroom!

In its simplest form, we have students talking with one another. We plan for and provide bonafide discussions for the students on the subject matter at hand. It is sometimes hard work to lead students in discussions that include *all* students. It is also difficult to remain on target. It can be done by a teacher who is willing to ferret out and practice the skills she needs to lead such discussions. There are plenty of resource materials around for learning those skills. A quick suggestion here is to provide the physical arrangements that are conducive to discussion activity. One component is the circular arrangement of the students' chairs/desks. Some teachers forget bolts were removed from the desks a half century ago. They insist on having student discussions—even though the desks are arranged in five rows of seven chairs, forcing students to look at the backs of one another's heads.

One of the clear findings we are getting from the research is that students, especially young students, need to talk about what they are learning. Young children need to develop their verbal skills, so they can talk about, describe, and capture for themselves the essence of what they have learned. They need others to talk to, to "set" that dialogue in the brain. If we had a dime for each "Shhhh" that was perpetrated on children preventing worthwhile dialogue in the classroom (and in the library), we'd be rich. Of course, we'd never want to get rich at the expense of children and learning.

Reciprocity among students and reciprocity between the teacher and students provides an extra motivational edge. Let's quickly point out that we do not advocate wholesale, unbridled talking. To the contrary. We do want to provide appropriate, ample, opportunity for dialogue to occur.

Reciprocity can occur through classroom discussions, as already mentioned. It can occur in committee work, in tutorial arrangements, in effective question and answer sessions, over a private lunch with one or more students, in debates, in open forums, etc. You can add to the list.

In Order for Students to Learn Fully and to Retain What They Have Learned, Students Must Engage in the Give-And-Take of Communication, Not Only with the Teacher, but also with Appropriate Others.

Reciprocity as an intrinsic motivator fares best whenever the teacher has developed a community for learning. Such a community requires a cohesive group of students and teachers working toward the achievement of common goals.

Summary

In summary, the five components of intrinsic motivation identified and explained in this chapter may be understood as concepts in their own right. The teacher can become skilled in activating them singly and by doing so can attain a measure of success in developing self-motivated students.

The five concepts are not mutually exclusive. They are best understood and used on an integrated basis. For example, the teacher would do well to provide for productive *reciprocity* among her students, that is, she could get them communicating with one another about the learning task at hand. In doing so, certain students would likely identify with certain others. The identification striving could be coupled to the striving to attain the *competence* the more able students may demonstrate. If the reciprocity goes well, an air of excitement or enthusiasm may be generated which is, of course, a form of positive *tension* or a t-state. The integrated effects of these intrinsic motives provide the best impetus for the student to learn. They "connect" well, one to the other.

Extrinsic and intrinsic motivational theory are available to the teacher. There is a crucial choice to be made between the two. Choosing the extrinsic motivational strategies will relegate the teacher to the role of dispenser of rewards and punishments, not a very attractive role. Choosing the intrinsic strategies will allow the teacher to take on the role of influencer. This is the better of the roles, for influence is the essence of the teacher at work.

In the final analysis, learning comes as a result of what goes on *inside* the student. If we could eavesdrop on the internal dialogue of the student we would surely know and understand several things: we are observing an individual who has a great deal of curiosity, among other emotions; one who wants to be competent, both in the physical and the social world; one who wants to take on certain characteristics of others and, in so doing, identifies with them; one who wants to experience meaningful relations with others through a striving for social reciprocity. These wants and strivings provide teachers with the greatest inroads to student motivation.

Intrinsic motivation sustains the learner when the teacher is not around. Its force engenders the life-long learner. May the force be with you the teacher!

Chapter 3

THE OBJECTIVE

Now that we've focused on the need for effective teaching and know that we've made a case for effective motivation of students, let's turn our attention to another critical area. That area is the student objective.

Effective Teaching Cannot Take Place Unless the Student Has Objectives to Achieve.

Objectives must be set for students. It is foolhardy to begin instructional activity without students having the *direction* that objectives provide. Too many times teachers begin their instruction without objectives and as a result the instruction becomes erratic and inefficient.

Another error occurs whenever the teacher does not inform the students what the objective is. The lesson begins without students having the benefit of knowing where they are going. It almost seems ridiculous to mention this error of omission, yet it must be mentioned. The error is made more often than not.

Students need to know where their lesson is heading. They accomplish more and they do it in less time when they know their objective(s). It's so simple and it takes so little time to inform the students. Some teachers simply write the objective on the chalkboard before the lesson begins and make reference to it at the most opportune time.

We will use this chapter as a means of identifying the fundamentals of sound objective development for the teacher. Since this is a book which deals primarily with the act of teaching, we will only speak briefly about the preparation of objectives. We need to address the topic of objectives because without objectives, the act of teaching would be meaningless.

Developing objectives, announcing the objectives to students and using teaching skills that will aid students in accomplishing those objectives are three steps teachers must make.

The following is the objective for you in this chapter:

Chapter Objective

Know the fundamentals for preparing effective student objectives.

First of all, the objective must be conceived and written with the student in mind. Study the following sentence that was conceived and intended as an objective: to present students with data supporting the fact that India has an overpopulation problem. You will note that this is really not an objective. At least it is not an objective as we are defining it. It directs the teacher, not the student. The key words that prompt us to arrive at that conclusion are the first two, "To present. . . . " The teacher will be doing the presenting. We want the emphasis of the objective placed on the student's behavior.

For the above procedure to be rewritten into the form of an objective, we will first say to ourselves the words "The student will . . ." and then think of an action word, a transitive verb, if you will, that describes what the student will be doing. The verb should depict what the student will be doing to verify the accomplishment of the objective.

EXAMPLE: ("The student will") Analyze the data presented by the teacher that supports the fact that India has an overpopulation problem.

We also like the verb stated as the first word in the objective because it sets an action tone to the objective. It helps give to instructional activity the *direction* of which we spoke earlier.

The grammarian would tell us that a transitive verb is action oriented; it takes an object. Action directed to an object, a target! That's what we want in teaching, student action directed to some learning target.

EXAMPLE: Pour one cc of H_2SO_4 into a 20 cc beaker containing one cc of H_2O.

The action is the pouring; the object of the pouring is the sulphuric acid. In this example there is even a physical danger for the student if the instruction is not in precise accord with the objective. If, for example, the water were poured into the sulphuric acid due to inappropriate instruction, a real explosion would occur.

Begin the objective with an action verb, one that describes the action the student will be taking. Another thing to determine is in which learning domain the objective will be conceived, the cognitive, affective or psychomotor.

Simply put, cognitive learning includes the thinking process; affective learning includes emotional activity like habits, attitudes and values; psychomotor learning includes physical activity like running, drawing, typing and writing. These domains have been thoroughly analyzed and published in a wealth of books. The basic books you may wish to consult are listed in the Bibliography.

We will give several examples in the most commonly used domain, the cognitive. There are six basic categories in this domain; knowledge, comprehension, application, analysis, synthesis, and evaluation. Each category is subdivided to varying degrees. The basic divisions, as well as the subdivisions, provide the teacher with a wealth of verbs as well as thinking activity.

Some Examples

(The student will . . .)

(1) *Know* the multiplication table of 3×1 through 3×10.

(2) *Comprehend* the difference between empathy and sympathy.

(3) *Apply* the formula $S = \frac{1}{2} gt^2$ to appropriate workbook problems on pages 55–58.

(4) *Analyze* the six mysterious ingredients in a shoe box, which will be provided by the teacher, and identify them as the parts that go together to make a clothes pin.

(5) *Synthesize* into a unique "contraption" six mysterious ingredients in a shoe box, which will be provided by the teacher.

(6) *Evaluate* United States involvement in Latin American countries these past twelve months; the evaluation criteria to be used are positive world image and favorable balance of trade.

The above examples are given for the basic six categories of the cognitive domain. As mentioned earlier, there are further divisions of each category. As an illustration, the category of comprehension has three divisions: translation, interpretation, and extrapolation. The action verbs would be translate, interpret, and extrapolate. Of course, synonyms can be used for these and any of the other verbs generated by the domains.

Let's look at a few examples of the subdivision.

Some Examples

(1) Translate the sentence, "Plus ça change, plus la même chose."

(2) Translate into his or her own words Shakespeare's soliloquy, "All the world's a stage. . . . "

(3) Interpret the advice, "Don't take candy from strangers."

(4) Extrapolate missing numbers in sequences such as 1, 2, 3, _____, 5; 4, 5, _____, 7; 2, _____, 4, 5; _____, 8, 9, 10.

(5) Predict (extrapolate) the direction the following graph line will take after point x.

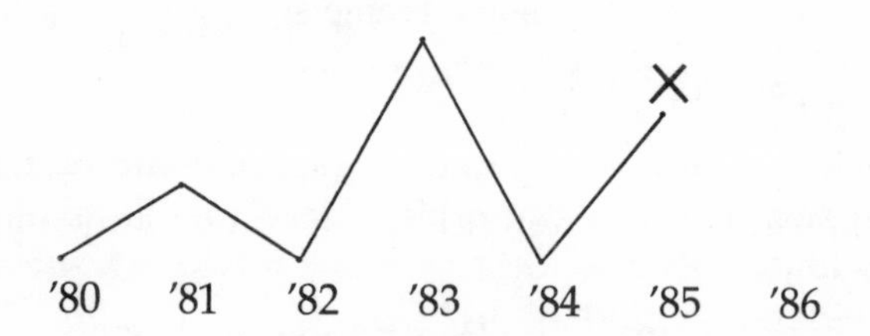

We'll leave the cognitive domain and give two examples each for the affective and psychomotor domains.

Examples (Affective)

(1) Develop respect for the other person's point of view.

(2) Value the idea that education is more than just preparation for college or a job.

Examples (Psychomotor)

(1) Execute the forward somersault, ending the maneuver in a standing position.
(2) Stir-fry simultaneously three different vegetables that will complement the fish sauce recipe learned in the previous lesson.

Let's summarize what's been said to this point about the fundamentals of objective writing.

(1) The objective is to be written in terms of what the student will do.
(2) The objective will begin with a transitive verb in order to give direction to the instructional procedures which follow.
(3) Objectives can be written in three learning domains, cognitive, affective, psychomotor.

A point that needs to be emphasized greatly will be our fourth point.

(4) Direct all instructional activity toward the accomplishment of the objective(s).

You'd be surprised how much instructional activity is extraneous to the objectives at hand. Teachers get "off the track," it is no sin to do so if it is done for sound reasons. Off the track means the instruction is not directed to helping students accomplish their objectives. That's what teaching is all about—helping students accomplish their objectives!

So much has been written in the professional journals and education textbooks on the topic "Time On Task," it seems unnecessary to discuss that topic here. Yet teaching to the objective is the very essence of time *on task*.

Researchers have found that as time on task increases so does student achievement increase. We are not talking about the amount of time spent on student objectives, which certainly is a worthy topic. We are talking about the second part of the researchers' point—ON TASK. On task means the students are working on the accomplishment of their objectives. Instructional activity must help that process.

Here is a fifth and final point about objectives. This point will clarify the major difference between a goal and an objective.

(5) Objectives are to be accomplished within a single period of time, goals take more time than one period.

In Chapter 2 we saw that striving for competence was a powerful intrinsic motive, one that the teacher would do well to elicit. When an objective is accomplished by the student, the teacher should make it known to the student that he is now competent in yet another area. To the learner, achieving is self-motivating. The teacher's role is to make sure the learner knows that he is becoming even more competent. That is a good point on which to end this brief chapter on the preparation of objectives.

Let's summarize. Prepare an objective for students which can be accomplished in a given lesson. Inform them what the objective is. Use appropriate teaching skills, activities and procedures that will help students accomplish the objective. Tell them when they have become competent at doing what the objective required. That's it—alpha to omega—soup to nuts! Sounds simple. It is. Sounds easy. It is not! It is very hard (but rewarding) work, work that can only be done by you, the professional educator.

Now we move on to eight chapters which will identify, explain and exemplify teaching skills that are critical in the action setting of the classroom.

Chapter 4

SET INDUCTION

The roll is checked. The absentee form has been sent to the office. The teacher is ready to do her first teaching of the morning.

She clears her throat, "Uhhhmmpff!"

The students glance her way.

"Open your books to page 36. Read pages 36 and 37. You may begin now."

So it goes . . . in thousands of classrooms, plunging right into the lesson, the assignment, the project, with little or no effort made by the teacher to get students needing or wanting to learn what's about to follow.

Think what it would be like to be a student in that class for that particular lesson. Have you ever been a student in a class that began that way? As a teacher, have you ever taught a lesson by starting out that way? You probably have. Most of us have.

All too frequently that's the way it is. A plunge beginning does not assure productivity in learning.

When Teachers Take the Time and Make the Effort to Begin Lessons Effectively They Are Making Sure There Will Be a Maximum Benefit in Student Learning.

This chapter will be used for the identification and explanation of the teaching skill that is designed for beginning lessons effectively. A few examples will be provided as well. The beginning of a lesson ought to provide maximum benefit in student learning. The skill we use to do this is called set induction.

Set has to do with psychological readiness in the student for the lesson that is about to begin. *Induction* has to do with activity in which the *teacher* engages for insuring the set in the student. In effect, the teacher helps induce the set in the student, but it is the student who decides whether or not he will allow himself to be influenced by the teacher's activity. By engaging in certain behaviors, the teacher increases the probability that the student will decide to be influenced to learn, that is, motivated to learn.

Let's look at the complete term *"set induction."* Set induction is best understood by analyzing the term as if it were a pair of verbs. Verbs put ideas in action form and that's what this teaching skill is—TEACHER ACTION! (The same is true for all the other skills.) Teaching skills are action-oriented and the teacher is the action-taker.

Thinking of induction in the infinitive verb form, the teacher has to induce. The teacher's job is to induce in the student a need state, a want state for the learning that will be available in the activity about to unfold.

When the student is influenced by the teacher's action to set himself for the lesson, he is psychologically ready and willing to learn. In this sense, to be set is to *want* to learn. That's the heart of set induction—the teacher has to get the student to *want* to learn.

Think about the old saying, "You can lead a horse to water, but you can't make him drink." Think of the horse as the student and the water as learning. If we made the saying go something like, "the teacher's job is to lead the horse to water and get him to *want* to drink," we have a clearer picture of what the teacher must do.

Want is a word that comes from the affective domain of human behavior. Affective behaviors include emotions, habits, attitudes, values, ideals, etc. The affective domain is where the teacher's set inducing actions are best directed. The teacher must tap into the emotional makeup of the student, if the student is to become motivated.

There are numerous student emotions the teacher can tap. To name a few, the teacher could work to activate a student's amusement, excitement, joy, curiosity, puzzlement, inspiration, empathy, awe, fascination, temptation, and wonderment.

When students experience these emotions they want to act on them. They *DO* something as a result of feeling what they feel. What they do, if directed properly by the teacher, will lead them to learn what the teacher has planned.

Looking at the type of emotions listed above, one might argue the case that things like curiosity, puzzlement, and wonderment are cognitive phenomena. So they are, but they are inextricably bound to the affective domain. The teacher reaches the student through either avenue, cognitive or affective, but the better, more effective route is through the affective. Remember, as teachers, we must get students wanting what's about to be taught and want is an affective consideration.

We need to make one further point about the purpose of set induction as a teaching skill. The teacher's use of set induction must evoke a degree of tension in the learner. Let's quickly understand two things: number one, tension is a prerequisite to learning, and number two, a moderate degree of tension is desirable if students are to become actively involved in the learning task.

A brief explanation of what we mean by tension may be in order. Tension, as used here, is any type of emotional state in which the student puts himself. If the student is in a state of fascination, wonderment, amusement, etc., we would see that as a form of tension. However, we would not like the emotion to be too strong, that is, beyond moderate. Strong emotions can squeeze or narrow the student's attention and effort to a point where concentration is put more on the satisfying of the emotion rather than on the learning of the task at hand.

Max Wertheimer, eminent psychologist, speaks of "t-states." Note the small letter "t." It is indicative of moderate tension. Students who go beyond moderate degrees of tension become overly curious, overly awed, overly puzzled and as a result act only to relieve the feeling rather than to relieve the feeling by learning the task at hand. Adding to Wertheimer's concept, we might call the overkill emotion generated a "T-state." A further point is that T-states can and often do lead to anxiety in students, and that is certainly counter-productive. What we're after for students is a t-state, not a T-state.

Let's look at our objectives for this chapter.

Chapter Objectives

Understand the definition and purpose of the teaching skill called *set induction*. (You've already accomplished this objective because you understand that set induction's *definition* is the preliminary activity in which the teacher engages to induce the student to a state of wanting to learn. The *purpose* of the preliminary activity, the set induction, is to ensure a maximum payoff in student learning.)

Understand the four criteria of effective set induction.

THE FOUR CRITERIA OF SET INDUCTION

Interest

Generally speaking, the teacher's first order of business is to get the student interested in what the teacher does just before he or she begins the lesson proper. In that sense set induction is an initiating activity.

At the heart of set induction are students *wanting* to learn. Wanting to learn is an emotional consideration. So too is student interest. Clearly the best pathway to getting students emotional about their learning is the teacher doing something that will generate interest.

At the beginning of a lesson or a major shift in the lesson, suppose the teacher reaches under her desk, pulls out a brown shopping bag and says, "Wait 'til you see what I have in this bag, you'll never believe it!" There are few, if any, students who wouldn't have their attention drawn to the remark and the physical presence of the shopping bag. Attention would soon give way to curiosity. And curiosity is what the teacher wants her students to experience.

Once curious, the students will "move" to satisfy their curiosity. That movement, of course, will be led by the teacher in the direction of the learning task at hand. The learning task will have been identified in the objective that was planned for the lesson.

The teacher must ask himself in his planning of the lesson what he will do to capture students' interest. After he gets their interest, he must then do something more to tap into other things, things like puzzlement, challenge, curiosity, amusement, inspiration, etc.

A good emotion to tap after student interest is attracted is curiosity. Curiosity is a very strong, natural motivator. It is an intrinsic motivator, so strong, so readily educed by the teacher, that we can say confidently that it is almost the prototype of student motivation. In first attempts at using a set induction, teachers may want to rely on curiosity as a "sure thing" before going on to the other emotions that will get students caught up in their lessons.

INTEREST

Amused	Awed	Brave	Capable	Challenged	Competitive	Curious
Delighted	Determined	Dubious	Eager	Electrified	Enchanged	
Excited	Fascinated	Glad	Gratified	Happy	Helpful	Honored
Impressed	Inspired	Joyous	Kind	Longing	Nice	Opposed
Pleased	Pressured	Proud	Puzzled	Refreshed	Reverent	Rewarded
Shocked	Skeptical	Solemn	Startled	Stunned	Sympathetic	
Tempted	Tense (mildly)	Unsettled	Uneasy	Wonderful	Zealous	

As for the different emotions teachers might want to induce in the learner, the box above gives some ideas.

Connection

Whatever the initial activity the teacher uses, there must be some connection between it and the lesson about to unfold. As an example, the mere telling of a humorous story or joke, while getting students to enjoy what's going on, does not satisfy this second criterion.

Suppose the set induction is designed to get students to want to learn a lesson on the value of cooperation. The teacher reads an inspirational passage from some book. It really inspires the students. Yet it has no connection to the topic of cooperation. Because there is no link, the criterion of connection has not been met.

Let's look at this criterion by way of a diagram. If the lesson is to take forty minutes and the planned set induction two minutes we see

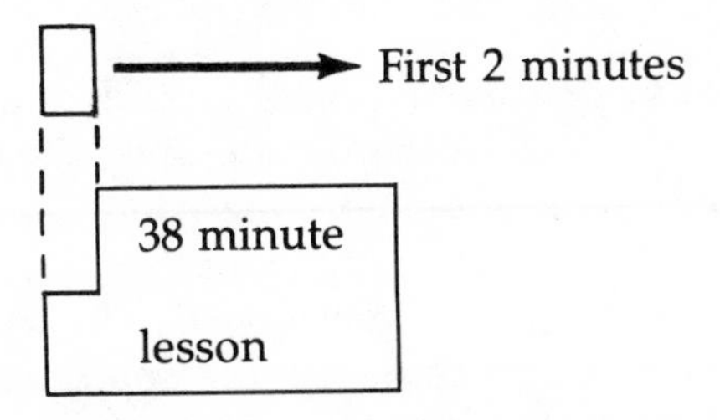

What the teacher does in that first two minutes must be relevant to the lesson. The student must see and feel that relevance by knowing what the connection is between the opening activity and

the main lesson. This connection is sometimes called the cognitive link. In that sense, the set induction must "hook" the student on what's about to come in the lesson.

Often teachers will tell a funny story or engage in some sort of irrelevant chitchat with their students as a way of relaxing them or removing a counter-productive tension. In itself, that may be a sound practice, but it is not a part of the set induction skill. There must be a connection between what precedes the lesson and the lesson proper.

The connection criterion of set induction serves one more critical function. It is the means for *setting* students ON TASK. On task means students working to attain the objectives of the lesson. The teacher has the responsibility of seeing to it that students know the lesson objectives and how the set induction activity is connected to those objectives.

Understanding

Almost always, the student must understand what is going on in the set induction activity. It is a case of *almost* always, because there may be certain times when the teacher intends to puzzle the students, inject a note of mystery, even confuse the students. Doing that, but not to a severe extent, can awaken intrinsic student motivation. Properly perplexed, the student will work to become unpuzzled, unmystified and unconfused. This mental "movement" can do a great deal for the student to master the task at hand.

The usual practice, of course, is to do something that the student does understand. If teachers are effective with this criterion of set induction, they should see the "light go on" in the faces of their students.

A critical part of *any* phase of teaching is student understanding of what's being taught. Without understanding, the student is left with the burdensome task of memorizing what was said, read, or done. The memorization is usually done with the thought, "I might have to know this for a test question." And if the stuff was on a test, the student retrieved the information from his memory bank and promptly began the process of forgetting (some psychologists refer to memorization as the first step toward forgetting).

Understanding as a criterion of set induction carries with it an extra measure of importance. The information provided in the SI activity serves as a building block for later information and skills that are presented in the main lesson. Effective teachers will often refer back to what was done in the SI. They make comments like "You'll remember what we did at the beginning of today's lesson. Well, this idea we're working on ties into that earlier idea you thought was so neat." OR "Now, this point is just like the one we made earlier, the only difference is . . . " OR "Do you see how this skill we're working on connects to the first one we did?"

Saying all this another way, the effective teacher connects the unknown to the known. What the student is about to learn is the unknown and what the student understood in the earlier SI is the known.

Memory

This is the simplest of the four SI criteria. All the teacher need do is make sure that whatever was done in the SI activity will be remembered for the duration of the day's lesson. That typically is a twenty to forty-five minute consideration and presents the teacher with no real problem.

There are times, however, when what the teacher uses as a set induction is so striking, so moving that it is remembered long after much of the rest of the lesson is forgotten. A few recollections of some inspirational set induction activities may illustrate the point.

Mr. McGlumphy, geography teacher, checks the roll, pauses, looks everyone square in the eye, each one in turn, makes a sudden move to the front left corner of the room. He grabs the world globe from its cradle, hefts it to the rear of his right shoulder, leans forward and says, "Quick, who am I?" After some playful and some serious student guessing he elicits Atlas as the correct response. He then launches into the background of how the ancient Greeks saw the world resting on Atlas' shoulder. He built some solid ideas on that startling beginning.

Then there was Mrs. McKee, English teacher, who in her quiet enthusiasm, titillated her students with a vocabulary that was so expressive her words almost jumped into their minds. One such experience involved her suggestion that her students concentrate on the color yellow. "Imagine a yellow so glowing, so irridescent that it makes one squint while looking at it," she said. "Imagine a yellow so golden that when viewed it prompts a reaction of warmth, of love." After fifteen to twenty seconds of concentration she gently tells her students how a father crazed by his earlier greed for gold begins to look at his daughter and her golden yellow hair in a new light. The teacher related that this was "how Silas saw his daughter, Effie, and the glow that radiated from her golden hair and her countenance was much like what you just experienced." The unknown of Silas' feelings is linked to the known of the student's private feeling.

And Mrs. LeFevre who "implored" her students *not* to read page 17 of the fifth grade grammar book. That page had a naughty work on it that "somehow got past the publisher" and it was "a shame that such a word had to be there for everyone to see—so please don't read THAT page." Needless to say, it was the most well read page in the entire book. Of course the objectionable word never was found but it certainly was not due to lack of effort. (There are some possible negative side effects of such a tactic, but the point here is the strength of the SI. What she did was remembered long after the grammar lesson was forgotten.)

All four criteria must be present for set induction to be effective. Once practiced, they are not difficult to manage and make a part of SI activity. Getting students *interested* in the lesson, exposing them to a beginning activity that *connects* or leads into the main lesson and making sure the students *understand* and *remember* what was done in the SI seem like common sense and basic ideas for good teaching. They are.

Initiating the act of teaching without a prepared sound beginning, a beginning that will enhance student learning, is to teach foolishly. Teaching without the use of set induction is just that—foolish, yet it is the second most serious error of omission that teachers make. The failure to use the skill of closure is first. Chapter eleven will be devoted to that particular skill.

In the next chapter we will look at a number of set induction activities. They are ones actually used by practicing teachers. A note of encouragement may be in order here.

Sometimes teachers feel overwhelmed by the realization that set induction is to be used for 180 school days a year times the number of different subjects taught. On top of that, set induction needs to be used at major shifts within a single lesson! The only advice that can legitimately be given is that teachers must think and plan for effective beginnings. What will they do to get the students to *want* to learn what's about to be taught? Teachers inevitably become amazed at how creative and prolific their ideas become once they work at it consistently, once they do it day in and day out, lesson in and lesson out.

The 10,000 mile journey truly begins with the first step!

Let's look at the steps taken by twenty-four effective teachers. . . .

Chapter 5

SET INDUCTION: EXAMPLES

In this chapter, you will read what practicing teachers used as set induction activities to get their students to want to learn the lessons about to be taught. One teacher's set induction may not suit another. Teaching styles vary, so do teaching personalities. As you consider using the following ideas, adaptations most certainly can and should be made.

Chapter Objective

Make a commitment to adapt and use some of the set induction activities that appear in this chapter.

Elementary

SI #1 TEACHER: Joan A. Jones

Grade Level	*Subject*	*Use*	*No. Students*
Kindergarten	Art	Introduce a lesson	23

Topic: Science and The Sense of Hearing

Emotion Tapped: CURIOSITY

DESCRIPTION

After the children had finished eating their snack, the teacher attracted their attention to something she began to do. At first, she said nothing to them, all the while making a bit of a show by pulling a large black box onto a rug where the children would be assembling. At her invitation, the children began walking around the box. Some sat on it, others moved it around. In answer to the teacher's question about the sounds coming from the box, the children discussed, for a brief time, the variety of the sounds. They tried to distinguish one sound from another. The teacher did not have to question them at length what the lesson for today was about. At the height of the guessing of the sounds coming from the box the class began calling out, "We know! We know! Our lesson is about what we hear!"

COMMENT

The teacher began the curiosity-evoking process by first attracting her pupils' attention, which is the beginning of curiosity. She did this by a silent, yet highly overt show of dragging out the big black box. Once the box was in full view of everyone, she got the children into the activity further by physical involvement (walking around the box, sitting on it, etc.). Then, with a well-phrased question, she asked them to guess what the sounds were. What they saw and heard certainly had them curious. What they guessed the sounds were prompted even greater curiosity—now they wanted to know if their guesses were correct.

SI #2 TEACHER: Cindy Snyder

Grade Level	*Subject*	*Use*	*No. Students*
Pre-First‡	Self-Discipline	To establish quiet when needed	16

Topic: Magic Box

Emotions Tapped: ENJOYMENT AND DESIRE TO COOPERATE

DESCRIPTION

One day during the first week of school, the teacher placed a white box on a table where the children would see it as they en-

‡Pre-First grade is a half-step promotion between kindergarten and first grade. It is designed for children who are capable of learning well, but who are not quite ready for the first grade experience. They are assigned to the Pre-First class for the advantage of time, maturation, and a caring teacher. The number is kept at sixteen so that personalized help is more readily available.

tered the room first thing in the morning. All morning long the teacher told them to stay away from this breakable, magic box. If the pupils talked out during that morning or they misbehaved in some way, they were told that the white box probably didn't like this behavior. When nap time arrived at noon, it was time to open the box.

Inside was a music box with two carved figures of children sitting on a seesaw. The music of the box was "Jack and Jill." After it was played for the pupils, they discussed how it made them feel, etc. They arrived at the decision that anytime the class heard the music they would become and stay very quiet and enjoy the lullaby.

In Miss Snyder's words, "The box works! When we do seat work the music box is played. We work quietly and do good work. When we nap, we do not move. We simply nap and listen."

COMMENT

With this SI we find the teacher working to get her pupils to cooperate through the relaxing effects of enjoyable music. She has kept herself out of the scolding, reminding, and nagging business some teachers get into. The soothing, calming benefits of enjoyable sounds helped instill the desire to cooperate, a strong set induction for whatever follows!

SI #3 TEACHER: Lisa Patrizi

Grade Level	*Subject*	*Use*	*No. Students*
1	Creative Art	Introduce a lesson	25

Topic: Creative Uses for Scissors

Emotions Tapped: SURPRISE, AMUSEMENT

DESCRIPTION

The teacher began the lesson by moving about the room opening pupils' desks as though she were looking for something. After several attempts she found a pair of scissors. Holding them for all the class to see, she said, "Oh, here it is!"

The pupils appeared surprised and curious in regard to the teacher's behavior. Mrs. Patrizi then asked, "What is this?" The pupils were amazed their teacher didn't know what she was holding, so they gave out a loud reminder, "It's a scissor [sic]!"

"Oh, really?" Mrs. Patrizi said. Holding the scissors against the chalkboard, she then began to outline the shape of the scissors with a piece of chalk. Completing that, she added some lines to the drawing to create the picture of a clown. Now, she asked again, "What is it?" The pupils laughed and responded, "A clown!"

Adding some more lines to a second scissors outline turned it into a race car. Without being asked, the pupils said in chorus, "It's a car!"

The teacher held up the actual scissors and said, "And you thought it was just a scissors!" The pupils giggled with delight and all became eager to outline their scissors and turn it into something of their very own. When they completed their work, Mrs. Patrizi said she was amazed to see a tree, a shark, a bird, a monster, a rocket, and even a bridesmaid!

COMMENT

Creativity, originality of idea and design, enthusiasm, pride in accomplishment, and more were the result of the SI this teacher used.

SI #4 TEACHER: Kimberly Cooke

Grade Level	*Subject*	*Use*	*No. Students*
2	Health/History	Introduce a lesson	20

Topic: The Four Food Groups and A Balanced Meal

Emotion Tapped: ANTICIPATION, CURIOSITY, EXCITEMENT

DESCRIPTION

This lesson is planned for implementation on or near George Washington's birthday. The teacher begins by having the pupils recall what special day it is (will be soon). Establishing this, the teacher then reads the booklet, *George Washington's Breakfast* (see bibliography section, Jean Fritz). Having heard this account, the pupils learned that Washington enjoyed "hoe cakes" as his breakfast habit.

Through some research activity, the pupils learned that a "hoe cake," or "Johnnie Cake," is nothing more than a cornmeal pancake. It was decided that on another day, the class would make its own hoe cakes.

COMMENT

You can see that Mrs. Cooke built up the interest level in the topic of eating through an analysis of what a famous person's meal typically included. Curiosity was built, then channeled into anticipation and excitement as the pupils began plans for the time when they would prepare what our first president had eaten.

SI #5 TEACHER: Margery Helm, Librarian

Grade Level	*Subject*	*Use*	*No. Students*
2	Library Science	Introduce a lesson	25

Topic: Selecting a non-fiction book

Emotion Tapped: AMAZEMENT

DESCRIPTION

Mrs. Helm believes it is important that students be aware of the variety and types of books which are available to them in the library. In this SI she assumed the role of a magician and with a lot of flair she showed the students a paper doll and a plain white envelope. She placed the doll in the envelope and told the students to repeat some concocted magical words. She said that with the help of students' good thoughts she would cut the envelope into two pieces without cutting the doll. Magic words having been said, she cut the envelope, deftly and secretly removed the doll whole from one of the envelope halves and showed it to them.

The students were amazed and wanted to know how she did that. Her reply was, "Magicians *never* share their secrets, but books *always* share their magic." She pointed out the location of the magic books in the non-fiction section. She encouraged her students to read and become magicians too.

COMMENT

You can certainly believe that this teacher/librarian provoked her students to learn through the emotion of amazement. (I was so intrigued by this idea that I even asked Mrs. Helm to send me the book in which she learned the envelope and doll trick. She did and I learned it. I use it with my students much to *their* amazement. It's an effective SI for me.)

SI #6 TEACHER: Joyce B. Zervanos

Grade Level	*Subject*	*Use*	*No. Students*
3	Reading, English	Introduce a lesson	16

Topic: Whisper Down the Alley (culminates in a writing activity)

Emotion Tapped: EXCITEMENT

DESCRIPTION

The teacher taught the students the game, Whisper Down the Alley, and had them play it a few times so they would understand better the kind of writing they would be doing. (The game involved whispering a message from one person to another until all students had it repeated to them. The last person hearing the message must relay the message he thinks the first person passed.) The students learned that there was a great deal of distortion due to ineffective listening. The teacher underscored the importance of being a good listener because listening is vital to learning to write well.

COMMENT

When students go through a game activity such as this, they will need help in making a valid connection with what the teacher wants done next. In the above case, it was creative writing. Once the connection is clear, there is typically an excitement generated for the main activity.

SI #7 TEACHER: Emma J. Baer

Grade Level	*Subject*	*Use*	*No. Students*
3	Art	Introduce a project	25

Topic: Architectural Drawing

Emotion Tapped: ENTHUSIASM

DESCRIPTION

This lesson was begun around Halloween when interest was high on "haunted houses." The teacher and students talked about and saw examples of the kind of houses used by civic groups, the motion picture industry, etc. as haunted houses. The students came to realize the houses are usually in a group known as Victorian. The discussion brought out the distinctive features of Victorian architecture. When they finished the discussion, Mrs. Baer showed how a small square of cardboard could be used as an artist's instrument for drawing a Victorian house.

COMMENT

Sometimes the obvious is obscured by the fact that we often look at things and yet we never really see them clearly. This art teacher's impact on her students was strong enough to generate an enthusiasm for the lesson for the day. She generated it by prompting her students to think about and to discern what makes certain houses different from all the rest. When they began seeing the differences, a natural enthusiasm occurred for taking their learning further.

SI #8 TEACHER: Linda L. Vickery

Grade Level	*Subject*	*Use*	*No. Students*
4	Science	Introduce a major topic once the lesson was begun	26

Topic: Adaptations of animals to their environment via body coverings

Emotion Tapped: CHALLENGE, CURIOSITY

DESCRIPTION

Miss Vickery provided her fourth graders the opportunity to examine different types of animal body coverings: fur, feathers, scales, skin. They were directed to make their examinations in terms of likenesses and differences. Prior to the lesson, the students had been asked to bring in pictures of animals. The teacher had a supply on hand in case some students had not brought theirs.

The teacher fastened four paper bags to the chalkboard. Each had a picture of a different animal covering on it. The students were told they would have to place their pictures in the bag according to the proper covering. It was identified as a classification task.

COMMENT

The children had really gotten ready for the main event of placing their animal pictures in the right bag. There was both a curiosity for how animals adapted to their environment and a challenge generated in applying their newly learned knowledge to the task at hand.

SI #9 TEACHER: Chris Snyder

Grade Level	*Subject*	*Use*	*No. Students*
5	Science	Introduce a lesson	23

Topic: Fossils

Emotion Tapped: CURIOSITY

DESCRIPTION

The teacher told the students a story about herself when she was a student in the seventh grade studying fossils. She recounted how she took a field trip to Deer Lake, where a construction project had left a gaping hole in a cliff. She drew a sketch of it on the chalkboard, firming up a more reliable picture in her students' minds. She then asked a curiosity-evoking question, "How could it possibly be that she found fossils of *marine* life in a cliff (of all places)!" The students acted to satisfy their curiosity by making some clever guesses. They soon realized that the area was once underwater and a lively discussion followed. Fossils that had been brought into the classroom were used to illustrate points being made.

COMMENT

In providing this SI example, Miss Snyder pointed out that ". . . While it is easy to deduce that this area (Deer Lake) was once underwater, it never failed to motivate the students, and there was always a lively discussion afterwards." This is seemingly a simple example of SI, yet there really is a lot to it. Here are three factors:

(1) The teacher related a personal story about herself, something with which students could readily identify.
(2) The teacher had brought to the classroom real fossils.
(3) In some further comments to me, Miss Snyder noted, "I think I can generalize and say that the asking of an interesting question which requires higher level thinking can really motivate." (Well said, Chris!)

SI #10 TEACHER: Joel B. Frantz

Grade Level	*Subject*	*Use*	*No. Students*
6	Health	Introduce a major topic after the lesson was begun	27

Topic: Structure and function of the heart

Emotion Tapped: CURIOSITY, PUZZLEMENT, CHALLENGE, COOPERATION

DESCRIPTION

(Let's use Mr. Frantz' actual words as he described his SI.) The following lesson concerns the structure and function of the heart and the path of blood through the heart. It is a review lesson. The students obtained adequate or thorough understanding of the heart before the lesson was presented.

"This lesson actually contains *two* set inductions in my opinion. The first set induction occurs the day before the lesson. The second set induction occurs the day of the lesson."

SET INDUCTION #1

"The day before the lesson I ask the students to move their desks to the perimeter of the classroom *(but I don't tell them why)*. This leaves a large space in the middle of the room. A simple thing like this generates curiosity and many students are wondering about the desk arrangement. In my opinion this tactic makes students wonder about and look forward to the next class.

"After school I use masking tape to diagram a large heart on the floor. The heart and connecting arteries and veins are large enough for students to walk through."

SET INDUCTION #2

"When students enter the room the next day I ask them to stand in the back of the room. They are viewing the heart upside down. At this point they have to guess what the diagram represents. Once I get the answer to that question I tell them they will be required to walk through the heart. I then ask them what they will represent (blood cells)."

COMMENT

Joel's words tell the story. I have observed this particular activity and it is a very powerful one. You can imagine that students are wholly immersed in the activity. They are captivated by it! He informed me that he has ". . . invented as many as five role playing activities that can be used with the 'giant' heart."

SI #11 TEACHER: James W. Gehman

Grade Level	*Subject*	*Use*	*No. Students*
1	Physical Education	Introduce a lesson	24

Topic: Ball skills (using a basketball)

Emotion Tapped: DESIRE TO MASTER A SKILL

DESCRIPTION

Here are Mr. Gehman's words: "Holding a basketball in my hand, I asked the class, 'What kind of ball is this?' Most first grade students answered 'basketball.' 'Yes, but how do you know?' was my response. By allowing the students to respond they were stimulated to *think*. I was looking for answers concerning size, distinctive lines, texture, a cover. Most students made reference to color which is not a unique characteristic. Holding another ball next to the basketball (e.g., soccerball) allows for visual comparison.

"This type of discussion stimulated the eagerness of the class to demonstrate to themselves and to me how well they can control the basketball as they 'dribble' the ball in a stationary position."

COMMENT

Knowing that a holistic development of the elementary age child is critical, this teacher tapped into all three learning domains: cognitive, affective, psychomotor. He used the affective one as the prime source for student motivation.

SI #12 TEACHER: Karen B. Orwig

Grade Level	*Subject*	*Use*	*No. Students*
4	General Music	Introduce a lesson	25

Topic: Writing parodies for well-known tunes

Emotion Tapped: DESIRE TO BE CREATIVE (in accord with the mood of the tune)

DESCRIPTION

Miss Orwig literally set the stage for this late Fall music lesson. She bound several four foot sticks of wood in such a fashion that when stood in a vertical position they formed a sort of pocket at the top. She placed some water and dry ice in a large black pot. The mix in the pot created a bubbling cauldron effect. She put the pot and its contents in the pocket of the sticks. Then she put a broom handle into the mix. She put on her witch's costume, replete with pointed hat and ugly wig and awaited the arrival of her class.

When the students were all in the room and totally captivated by this weird scene, she began stirring her "brew," all the while singing a song parody for "Jingle Bells."

Some students were taken aback, some giggled. All were caught up in this opening activity. ALL WANTED to see what would happen next. The teacher led them in a few paradies and then directed them to the lesson, namely, the development of parodies of their own. They were eager to do so.

COMMENTS

Not every teacher will want to don a witch's costume and stir water and dry ice to tap student's emotions for such a strong set induction. Too bad! A little zaniness is a great elixir—for teachers as well as students.

Secondary

SI #13 TEACHER: Daniel S. Witman

Grade Level	*Subject*	*Use*	*No. Students*
7	Industrial Arts	Introduce a lesson	15

Topic: The use of the forge

Emotion Tapped: ENTHUSIASM

DESCRIPTION

The teacher began the SI by asking if any of the students had ever visited Hopewell Village, Sturbridge Village or Williamsburg, Virginia (the first two are local historical sites where famous iron forges once operated). He immediately got an affirmative response from one-third of the class. The students were eager to relate their experiences. That eagerness, plus the teacher's enthusiasm for the topic, affected the remainder of the class so that they too became enthusiastic about the topic.

COMMENTS

Here you see a rather common approach for getting students into a topic, the teacher's own infectious enthusiasm. The fact that the SI approach is common should not minimize its importance. We all know how much of an impact an enthusiastic teacher made on us as students and how much more we wanted to learn whenever the teacher was enthusiastic.

SI #14 TEACHER: Albert D. Nye

Grade Level	*Subject*	*Use*	*No. Students*
7	Health	Introduce a lesson	25

Topic: Drug and alcohol abuse

Emotion Tapped: CURIOSITY

DESCRIPTION

Prior to the arrival of the students in the classroom, the teacher prepared a display of various drugs and charts. The display gave very vivid illustrations of the diseases associated with abuse. When the students entered the room they were immediately attracted to the display. Spontaneous discussion followed. The discussion was a natural lead-in to the lesson dealing with the harmful effects that drugs have on the body.

COMMENTS

Sometimes the teacher can say nothing and still generate interest and curiosity. In this case, the teacher did something in advance of the health period. He elicited the natural effect of curiosity from an eye-appealing display.

SI #15 TEACHER: Patricia A. Owens

Grade Level	*Subject*	*Use*	*No. Students*
7	Reading	Introduce a lesson	18

Topic: Six steps of a plan for remembering

Emotion Tapped: EMPATHY

DESCRIPTION

Mrs. Owens told a fictitious tale of a seventh grader, supposedly the son of a friend of hers. The boy had a real problem with forgetting. Personal but common examples were used such as his leaving his lunch at home on the table, leaving his science book on a desk in his bedroom, forgetting to do his homework, being late for Boy Scouts and running an errand but forgetting the items he was to purchase. Concern for the friend's son was expressed. Assistance from the group, using their personal experiences, was requested. The students eagerly contributed.

COMMENTS

The students naturally related to the problem set up by the teacher. They could readily empathize with the hapless son who had a problem of forgetting. Empathy is a valuable emotion for the teacher to elicit, not only for the benefits of intrinsic motivation inherent in the set induction, but also for the societal need of mutual understanding.

SI #16 TEACHER: Judith Williams

Grade Level	*Subject*	*Use*	*No. Students*
9	French II	Introduce a lesson	15

Topic: Words for various items of clothing

Emotion Tapped: AMUSEMENT

DESCRIPTION

An oversized bag concealed Miss Williams' clothing props for her set induction. The students were asked to guess (in French) the contents of the bag.

She randomly pulled each garment from the bag putting them on one at a time. While doing so, she named the French equivalents and students repeated them accordingly. After a while the teacher had donned a sizeable number of different items of apparel, making her get-up look a bit outlandish. When all items were on (over a dozen), she reversed the process returning each item to the bag as the students recalled the French equivalents.

COMMENTS

One can readily picture that setting and feel the amusement that must have taken place. That amusement, that emotion must have enhanced the students' desire to get into that lesson. If you were one of the students, wouldn't you?

SI #17 TEACHER: Steve G. Voguit

Grade Level	*Subject*	*Use*	*No. Students*
10	American History	Introduce a lesson	22

Topic: Assassination of Francis Ferdinand

Emotion Tapped: AMUSEMENT

DESCRIPTION

As the students were entering the classroom they noticed that Mr. Voguit was behaving in an unusual way. They also saw two words written prominently on the chalkboard, "Murphy's Law." The students watched the teacher drop folders full of papers, fumble for pens, hit his knee on an open desk drawer, etc. Then he asked if anyone could define Murphy's Law. Having arrived at a satisfactory definition, he pointed out the area of a resource book in which the assassination of Francis Ferdinand is covered. He then pointed out to the class the elements of Murphy's Law in the assassination. The students clearly saw the connection and understood the event in history.

COMMENTS

Here's an SI activity where the teacher engaged in a bit of theatrics that paid off. Students liked the dramatic effect that showed concretely how fouled up something so routine to the teacher as getting ready for his lesson could be. Through some laughter at the outset of the lesson students were drawn to where the lesson was going. In effect, they *wanted* to get into the lesson.

SI #18 TEACHER: Daniel R. Leininger

Grade Level	*Subject*	*Use*	*No. Students*
11	English	Introduce a major topic after a lesson was begun	22

Topic: Thoreau's "Civil Disobedience"

Emotion Tapped: CURIOSITY, DESIRE TO UNDERSTAND

DESCRIPTION

Let's use Mr. Leininger's comprehensive description of how his SI was planned and implemented.

Henry David Thoreau's *Walden* and "Civil Disobedience" are key elements in 11th Grade Advanced English. Both works contain many controversial ideas worthy of class discussion; however, there is a tendency for interest in Thoreau and his rebellious, critical point of view to wane (at least in the '80s) as the days pass. Therefore, I must constantly show students that Thoreau's ideas and observations, which are expressed in a ponderous, erudite 19th Century manner students find difficult to comprehend, are still relevant today.

Many students resent the fact that, in "Civil Disobedience," Thoreau dares to criticize the United States Government—their government! Some are ready to join the chorus of superficial readers who jump to the conclusion that Thoreau must be a Communist—how else could he criticize the best government, the best people on earth?

To help students through difficult passages, I supply a 29-question study guide. The first two days we discuss the first two-thirds of the questions, and generally put to rest their suspicions that Thoreau is a Communist; students become more open-minded by the time we have discussed the distinction between law and justice—then and now.

By the third day on "Civil Disobedience"—the one for which this set induction is intended—we are ready to discuss Thoreau's criticism of the rich (". . . the more money, the less virtuous . . ."). At the beginning of class, I place the following quotation on the board:

> ". . . The average American soberly regards money as the sole means of ensuring personal freedom and independence. This attitude towards money engenders an indifference to the means by which it is obtained."

We discuss the meaning of the quotation, and then discuss whether or not it is accurate, with particular emphasis on the meaning of freedom. Students have assumed the statement is from Thoreau, and are generally dazzled to discover it is taken from the preface of a KGB manual preparing undercover agents for work in the United States (the quotation appeared a number of years ago in a *Time Magazine* article on the KGB). Not only does valuable discussion follow on whether the Russians "really" understand us, but also there is an increasing awareness that an understanding of Thoreau's perceptions has validity today. Many students who agreed with the quotation on the board are now in the uncomfortable position of finding their opinion coincides not only with that of Thoreau (whom the teacher, obviously, is "pushing"), but with that of the Russian secret service as well!

Students are now mentally prepared to look at Thoreau's criticism of the wealthy in a new light—they have a personal stake in a resolution of the question. Many are now eager to see a flaw in the KGB position, and look for a flaw in Thoreau's position. In short, thinking is taking place, and at the very least, students are beginning to discern some gray between the black and white of opinions instead of blindly copying down "absolute answers" to questions on a study guide.

COMMENTS

Little needs to be said here. You can see how the teacher extended the SI beyond the usual first few minutes of the lesson as well as the minutes in a major shift. He actually used two class periods, 88 minutes, to get his students wanting to get into "Civil Disobedience."

SI #19 TEACHER: Donald B. Troutman

Grade Level	*Subject*	*Use*	*No. Students*
10 and 11	Advanced Algebra 2	Preparation for a test	25

Topic: Major test

Emotion Tapped: ENTHUSIASM, DESIRE TO COOPERATE

DESCRIPTION

Two days before a major test the teacher announced to his class that the upcoming test would be composed of some difficult problems. The students were told the test would be an open book test and that they may use calculators or small computers, as well as any other books for assistance. He added the interesting and exciting notion that the students would be working in threes. The students were told that each member of the triad would receive the identical grade so they ought to choose their partners wisely.

COMMENTS

"What enthusiasm!" Those were the teachers' words to me. He said the students really prepared and learned a great deal from that particular test, so effective was the SI! One student expressed the impact of the SI very clearly, "I learned more during the test than I did the last week of classes." In the last chapter it was mentioned that sometimes the set induction will be remembered long after the subject matter was forgotten. Mr. Troutman's SI is a case in point.

SI #20 TEACHER: Ann B. Worley

Grade Level	*Subject*	*Use*	*No. Students*
11 and 12	Home Economics Child Development	Introduce a lesson	20

Topic: Child's play

Emotion Tapped: EXCITEMENT, ANTICIPATION

DESCRIPTION

The class was divided into groups of five students each. All groups were provided with a tub of water, a sponge, detergent soap, an egg beater, a wooden spoon, some straws and a dish towel. The students were invited to play with the materials in whatever manner they chose. They were also given to understand that they were responsible for cleaning the classroom at the end of the play period. They were then given unrestricted freedom to explore water play.

COMMENTS

Here we see the teacher setting up a simulation exercise in which students would actually do an activity that they would later analyze young children doing. Because the students felt an excitement for doing this unusual simulation, they looked forward (anticipation) to the latter part of their lesson when they would analyze children's play using the same implements.

SI #21 TEACHER: Barbara H. Noble

Grade Level	*Subject*	*Use*	*No. Students*
12	Shorthand II	Introduce a lesson	15

Topic: Taking dictation

Emotion Tapped: CHALLENGE

DESCRIPTION

Prior to the students' arrival, the teacher wrote words in shorthand outline on the chalkboard. They were the words that would be used during the period's dictation on new material. The students were told to work together in "spelling the outlines" in order to master the unfamiliar, more difficult words. As the time drew nearer and nearer for the "live" dictation, the sense of challenge built to a productive point. The students became ready for that challenge.

COMMENTS

At first reading, this SI may not appear to be much, yet we note the commitment of the teacher to help the students with the tougher words. She lets them know she wants them to succeed and that comes through in how she organized the lesson flow. The SI is a valid beginning.

SI #22 TEACHER: William P. Moyer

Grade Level	*Subject*	*Use*	*No. Students*
12	Advanced Biology	Introduce a lesson	20

Topic: Color reception in the eye and after-image in the brain

Emotion Tapped: PUZZLEMENT

DESCRIPTION

Mr. Moyer had his students stare at a large red book which was placed on a table at the front of the room. Students stared for 45 seconds. Next they were directed to look at the light colored wall or the film screen positioned at the front of the room. The effect of the after-image in the brain caused students to see the outline of the book on the wall or screen, however, the after-image was green. The teacher then led the class in a discussion on the theoretical analysis of what caused both the after-image and the change in color.

COMMENTS

This was a very vivid and effective way to get students puzzled and then excited about finding answers to what puzzled them. Just reading about this activity sets up a puzzlement in my mind. And you?

SI #23 TEACHER: Carol M. Hoffman

Grade Level	*Subject*	*Use*	*No. Students*
7, 8, 9	Student Volunteer Group	For an assignment	24

Topic: Giving of oneself

Emotion Tapped: INSPIRATION, DESIRE TO HELP (EMPATHY)

DESCRIPTION

Having felt the need to tap the empathy potential in her junior high class of volunteers, Mrs. Hoffman began her lesson by asking, "What is a giver?" With no response from the students, she asked, "What is a receiver?" After a minute of silence, she asked, "Is it better to give than receive?" Silence continued along with nervous squirming and foot shuffling. After 30 more seconds of silence, she read *The Giving Tree* by Shel Silverstein. When she finished reading the book, the students were ready to think about giving and receiving. They were also ready to talk about it.

COMMENTS

Here we see a teacher who knew exactly how to get her students wanting to learn. She used an inspirational work, *The Giving Tree,* which is almost a sure-fire inspiration to those exposed to it. This was a critical lesson in the student volunteer program and the right beginning had to be used. Mrs. Hoffman's set induction was that beginning.

Summary

In this chapter we looked at a number of set induction examples. If you were induced to generate some ideas of your own *that you will use,* the chapter objective was accomplished.

I have had the privilege of observing all the teachers whose set induction activities were exemplified here. I even observed some of the SI's described. They were truly effective!

These teachers really know how to whet the educational appetites of students so they want to learn! Teachers who use set induction are teachers who are more effective in their work.

As a way of bringing closure to this chapter and solidifying your accomplishment of the objective, let's have you try an SI assignment. You're teacher #24 and you have to get your eighth graders wanting to discuss the effects of stress in a difficult job like the President of the United States. You have a pair of pictures of a rather well-known U.S. President. Here are the lesson particulars followed by the pictures. Let's see what you can create in the way of an idea for inducing an emotional set in your students that has them wanting to learn.

SI #24 TEACHER: You

Grade Level	*Subject*	*Use*	*No. Students*
8	American History	For a major topic after lesson was begun	25

Topic: Effects of job stress

Emotion Tapped: (As teacher, you decide)

Earlier in Presidency

Later in Presidency

Think of four (4) different ideas for getting students wanting to learn. Jot down your ideas below.

IDEA 1

IDEA 2

IDEA 3

IDEA 4

PART II

Sustaining the Middle of the Lesson

Chapter 6

Reinforcement Cueing

Teachers will find reinforcement cueing a natural skill, one that is part of every teacher's skill repertoire. They really have no choice about using it! The choice that teachers *do* have is how well they will use the skill.

Mastery of reinforcement cueing requires two things: an understanding of several principles inherent in the skill and appropriate practice. Let's go after the following objectives to help with your mastery.

Chapter Objectives

Understand each of the two basic categories of the reinforcement cueing skill, verbal and nonverbal.

Know a variety of verbal and nonverbal reinforcement cues for use in the teaching-learning process.

Understand the effect of reinforcement cueing on the learner.

First things first with this skill. Just what is reinforcement cueing? In answer to that question, let's say what we're not talking about.

We're not talking about the strengthening of what students have already learned. Strengthening knowledge and skills of students is accomplished by using teaching strategies that reinforce learning. Such strategies include reteaching already-learned material in a different way or drilling students on familiar material or taking students through a brisk simulation, such as a game of using learned material as the content of the game. Those activities *are* reinforcing to the learner and they are vital to the teaching-learning process, but they are not a part of the teaching skill known as reinforcement cueing.

Reinforcement cueing is teacher behavior, verbal or nonverbal, that typically occurs after a student has answered or asked a question or has done something related to the lesson presentation. We're talking about what the teacher says or does in the way of a "cue" to the learner. The cue informs the student of the correctness of his recitation or his performance. Students have come to expect such teacher behavior.

At the beginning of this chapter the statement was made that teachers have no choice of whether they will or won't use reinforcement cueing as a skill. That's true because teachers naturally react and respond, both verbally and nonverbally, to their students' behaviors. Those reactions and responses are read by the students as feedback to their performance.

Students naturally monitor their teachers' reactions and responses to their performance in the classroom. Students make adjustments to their performance, present and future, in accord with what they interpret. Saying it succinctly, students are greatly influenced by the way teachers behave toward them.

Reinforcement cueing occurs at the moment following the student's participation. That moment has great potential for effective teaching.

Let's say Becky responds brilliantly to a question you've just asked and it's her first comment in eight days. As her teacher you want to give her a cue that says you're thrilled with her answer because it was so good. So you let it all out, as the saying goes. You beam from ear to ear and you say, "Becky, what an answer! That was so well thought out I'm going to write it on the chalkboard" (and you do).

You've used several aspects of reinforcement cueing and you've effectively influenced your student in a positive way. Initially you've let her know her answer was correct, which is one of the purposes of reinforcement cueing. Next, you've let her know there was something about her answer that deserved special recognition for her, which is another purpose of reinforcement cueing.

Doing these two things for students, letting them know the correctness of their answers, comments, performances, etc., plus recognizing them for what they have contributed, ties in with the motivation forces that reside inside each student. And you've tapped the better kind of motivation, namely intrinsic motivation. (You will recall we discussed the importance of that type of motivation in Chapter 2.)

Students who find that what they say or do in class is positively recognized, even appreciated, are much more likely to increase their participation. That's what we want and that *is* the ultimate purpose of reinforcement cueing, to get the student to participate again and again, to stay active in the development of the lesson.

Active students are prone to learn more and learn better than those who are not. Of course there are a few students, very few, who learn best by being mentally active and in tune with the lesson while physically passive. We're not talking about them here.

You might ask at this point, "Well, what if a student answers my question incorrectly?" or "What if what the student says or does is improper?" You simply acknowledge the answer or behavior as wrong. Say, "Nope, that's not it" or "I'm afraid you're doing it wrong, Julie." Depending on the sensitivity of the student or what the effect negative cues might have upon the student, respond accordingly. Common sense prevails here.

Some educators advise teachers never to tell or show a student he is wrong lest they "bruise" the student's psyche. That hardly ever is the case. Students don't mind being informed of their wrong answer or performance. *How* the informing is done is the common sense key.

One can hardly observe Nick hitting Marsha on the head with his book and avoid saying, "Stop that right now, Nick," In this case, you don't concern yourself about bruising Nick's psyche. You must, however, concern yourself about potential bruises on Marsha's head. Dealing with that type of student requires disciplinary action and is dealt with more extensively in *Misbehavin'*.§

In summary, then, reinforcement cueing has two primary purposes. One is to let students know the correctness or incorrectness of what they have said or done. The other is to credit them, by way of recognition, for their contribution to the lesson. The longer term purpose is to get and to keep students *actively* participating in your lessons. By doing these things you are sustaining the lesson, the students and even yourself. You sustain yourself because it feels

§See Bibliography section.

good to spread sunshine over your students whenever you see them beginning to grow.

Think what else you've done when you said, "Becky, what an answer! That was so well thought out, I'm going to write it on the chalkboard" (and you do). You've indicated by your tone of voice and by your genuine smile that you're affected positively by what Becky said. We all know that most everyone likes to influence other people, especially in positive ways. That's what your smile and tone of voice communicated—that Becky is influential enough to influence you! And you're the teacher for Pete's sake! Hey, you even said you were going to write what she said on the chalkboard. You were going to take the time and trouble to do that! And did we mention that what she said would also contribute to the success of the lesson flow and that it would also influence what happened to Becky's peers, the other students? No, but we need to.

Acknowledging students' ideas and feelings on a subject, accepting what they say and what they feel, has a motivating effect on what they do in this lesson as well as future lessons. Research has borne out the fact that students who have teachers who accept and use student ideas and feelings as a part of lesson development achieve more than students whose teachers do not do these things. This important aspect of reinforcement cueing will be given greater attention in the next chapter.

Let's return to the effects of reinforcement cueing on the learner. We learned in Psy. 101 that behavior which is rewarded, from the learner's point of view, is more likely to recur. That's what reinforcement cueing does and more. We'll develop further the understanding of this skill, but for now it would be good to look at some examples of what teachers can use as verbal and nonverbal cues.

The typical verbal cues teachers use to reinforce student responses are ones like: Good, Right, All Right, Very Good, Correct, Excellent, OK, Fine.

We all use those cues and we use them so frequently they have become the least effective. Students have been overexposed to them. Such cues should be judiciously avoided in favor of others like:

"That's an effective thought."

"Sure, and that fits what Mark said a while ago."

"You've got it!"

"On the money!"

"Yes, Billie. You really have your thinking cap on this morning!"

You'll note that the two factors or primary purposes of reinforcement cueing are present in the above expressions. When students heard those cues they knew what they had just said or done was correct and they got the clear impression they were being recognized for their effort. The impression came from what the teacher emoted while cueing the students and from the unique expressions that were directed especially to them.

Let's talk about a problem here. Even when teachers use a variety of cues, even when teachers use unique expressions to cue their students, they often fall short on the need to personalize the cues.

In Order for the Reinforcement Cue to Have Maximum Effect, the Student Must Believe and Feel the Cue Is Directed Personally to Him or Her.

The simplest way to personalize the cue is to put the student's name in the cue. For example, you might say, "Wow, Jennie, that's a complicated question you just asked!" A person's name puts the person in the cue.

A more difficult yet more effective way to personalize reinforcement cues is by forming a cue that is appropriate to what the student has just said or done. For example, a student says emphatically, "Well I sure wouldn't pour the HCL in the water!" The teacher personalizes and says, "Not only is your answer on target, Mary, but the way you told us warns us of a potential problem. Class, what is the problem Mary was warning us about?"

Note the teacher picks up on the correctness of Mary's answer and on the special contribution she makes in her implicit warning about pouring acid into water.

Another example . . . A student skillfully performs a backward somersault. The teacher says, while giving a wink of his right eye, "I knew you could do it, Charlie. I'd give that move a ten." Note the recognition aspects inherent in the cue.

To put verbal reinforcement cueing in perspective, it can be said that the number of different cues is infinite. Striving for unique cues that are the most effective is a matter of the teacher's creativity and committed practice to the mastery of the skill. May I cue you at this point and say, "If there's anyone who can use cues effectively, it's you. I know you can do it, ____________________." (Wish I could call you by name.)

Remember, personalize those cues! Use the student's name or give a cue that fits the student or her response or do both!

One more thing before we get to *nonverbal* cueing. Do *not* cue every answer, every participation of the student. Such activity will tend to extinguish student participation. That is something you don't want. Too much of anything has an opposite, negative effect. (We learned this in Psy. 102, I think. Overreinforcing extinguishes the desired behavior.)

As for nonverbal reinforcement cueing, it is accurate to say that it is far more effective than verbal reinforcement cueing. Students, like all people, are more likely to believe what they see others *doing* rather than what they hear others *saying*.

Some communications experts claim that 80% of all that's communicated from one person to another is done through nonverbal channels. If that's so, then teachers would do well to master the several aspects of nonverbals, things like facial expressions, gestures, body language, tone of voice, etc. Those are the things that students tune in as they observe their teachers in action. Sure, they listen to what teachers say, but they really "listen" to what teachers do.

Let's consider some examples of nonverbal reinforcement cueing. Going to the face first, we have all manner of looks and expressions that can cue the student. There's the already-mentioned wink and the tone of voice that communicate effectively. Smiling, raising the eyebrows, opening the eyes wide, opening the mouth, looking interested, making eye contact and pursing the lips are some of the ways the face can be used to make expressions that reinforce student participation. Of course the expressions must communicate a positive reaction/response to the student. They must also be genuine and not forced.

Moving to hand gestures, there's the OK sign, thumbs up, the clapping of hands, raised fist, clasped hands over one shoulder, the touchdown sign football referees make, things like that.

Movements of the body as positive cues can be the affirmative nodding of the head, moving toward (or away from) the responding student in a supportive way, writing students' answers on the board and putting hands on hips to show pleasure with the student's performance.

Then there's the physical contact with the student, physical contact that lets students know they've done all right. Patting the shoulder or back, hugging, tousling hair, squeezing the arm, shaking hands, playful rap on the arm, contacts like that. Needless to say, when physical contact with students is made, care must be taken so that the contact cannot possibly be misunderstood. Common sense and sound judgment must be used.

We have looked at personalizing verbal cues. Now we need to look at personalizing nonverbal cues.

With a little thought you can readily understand that nonverbal cues are the epitome of personalization. How else can a student interpret solid eye contact with the teacher? The teacher's generous smile at him? A "thumbs-up" signaled his way? A tweak on his cheek? Those things are unmistakably personal and they make a positive impact on the learner.

One of the first things the teacher must do is harness all those nonverbal cues that are really a natural part of human interaction. The cues need to be harnessed, then practiced so that they can be used at will *as the situation dictates.* Nonverbal (and verbal) cues must be given judiciously, neither overdone nor underdone. The type student for whom the cue is intended is the key. Some students need more cueing, some less. The teacher is the judge.

For those teachers who are not so animated in their nonverbals, it is vital that they develop the animation. A Marcel Marceau school of the mime wouldn't be an outlandish idea for teacher preparation or inservice.

A word of caution here. With the concentrated effort to develop and practice nonverbal behaviors, it is conceivable that artificiality could creep into the behaviors. Sincerity is an important part of the cue. It's so important, that someone once said jokingly, "Be sincere, even if you have to fake it!"

Getting caught up in the act of teaching and especially in the growth of one's students is a sure-fire way to prevent insincerity. Teachers caught up in those two things will radiate genuineness as a positive reaction to student participation.

As teachers you will develop your own repertoire of verbal and nonverbal cues and those will, in all probability, become your best and most effective cues. Yet you will have to take care not to overwork a few cues so that you become predictable, thus losing spontaneity that is so vital to effective teaching.

Work for variety. Empathize with the reciter and think what she is thinking and feeling. Then say and do something that suits the student's thought or feeling. That will make it spontaneous, unique and personal.

One further point needs to be made about the two types of cues. Whenever a verbal and a nonverbal cue are simultaneously given, there must be a congruity between the two. Let's say the teacher makes eye contact with all students, one at a time, by glancing up and down the rows. All the while the teacher is wearing a frown. At the conclusion of the nonverbal activity the teacher says, "You're doing a fine job, class." Which do you think the students believe, the fine words or the frown?

Conflicting reinforcement messages can confuse the student. One thing is almost certain. Whenever students pick up conflicting cues they will invariably believe the nonverbal message, especially if it is a negative one.

Putting the two types of behaviors together for a combined cue is a teaching skill that can have a powerful impact on the learner. The cues must be both positive and congruent with one another. They must be sincerely given and they must relate to the student as a unique person. In that sense the cue must be personalized.

At this point we conclude our chapter. We have examined the basics of reinforcement cueing. There is more to know and understand, so we'll develop that material in Chapters seven and eight.

Chapter 7

Encouragement vs. Praise

We're sitting at a high school commencement ceremony. The speaker is repeating a familiar message. ". . . you're about to go out into the world and make your mark . . . you're on your own now . . . you've been wanting to be independent for a long time . . . you have arrived . . . this is it . . . you're finally independent!"

That speech and thousands like it are made across the United States June-in and June-out. The message is inaccurate and for a number of reasons. We'll deal with only one of the reasons—student dependence on the teacher in the instructional setting.

An example followed by an explanation will make the point. Picture in your mind a classroom. You're a student in that classroom and I'm the teacher. I ask you a question: "What does *heinous* mean?" You say it means odious, abominable and detestable. Then what do you do after you answer? Take a moment and think about your next thought.

You would no doubt look to me to see if I'm going to give you some sign that you're right or wrong. Turn this page upside down if you want to see whether *heinous* means odious, abominable and detestable.

Why are you doing this? Must you always look to the teacher to check out your answers? I'll bet you'd like some praise for your answer—if it was correct. Why not be an *independent* learner and look it up in your dictionary?

In classrooms all across this land teachers develop dependent learners. We do it unthinkingly. We do it unthinkingly because we reward student performance for correct answers, accurate home work papers, proper procedures and quality projects. We do it by praising the learner either for an answer or a product like a nice painting, a bird house or a quilted pillow. We also praise the student herself with a "You're a good girl, Wilma" or "You're an excellent student, Rosemary." And the students like it and they work harder for it. That's right! They work for it and "it" is the praise that comes from the teacher. Much time and space would have to be taken to elaborate on this situation. Let's just say at this point: all praise is not productive, in fact some is often counterproductive. It is counterproductive whenever students do their work just to get the good feeling associated with praise.

Let's take a moment and see what the objectives for this chapter are. We are already into the first objective. Hope you want to know more about it as well as three others.

Chapter Objectives

Understand the negative effects praise can have on the student.

Make a commitment to avoid the use of praise or at least reduce its use considerably.

Understand the positive effects encouragement will have on the student.

By continuing with the chapter's opening commentary on the negative side effects of praise, you ought to be sensing the need to avoid excessive and indiscriminate use of praise. In doing this you will be working toward the accomplishment of the first two objectives of this chapter.

You're probably asking, "You mean to tell me I'm not supposed to use praise for anything a student does?" Basically, that's what's being said, but then not using praise would be like asking you to take a cold shower. No one wants to do that. No one wants to stop anything "cold turkey." You're going to be using praise, that's certain, but learn to use it sparingly. Use it when it's absolutely needed to lift up a discouraged student. Then, when he's lifted up and can do without the praise, stop praising and go to encouragement as the basis for your ongoing teacher-student relationship. We'll deal with encouragement shortly, but let's take a further look at this idea of praising from the view of an expert in the field of human relations.

A few years back, Haim Ginott wrote an invaluable book entitled, *Teacher and Child.* In it he dealt with two types of praise, evaluative praise and appreciative praise. The chapter in the book that dealt with that dichotomy was aptly named the "The Perils of Praise."

Ginott wrote specifically about the destructiveness of evaluative praise. He made the case for its having a dependency-evoking effect on students.

The Essence of Quality Education Is Students Developing Self-Reliance, Self-Direction and Self-Control.

Those qualities require reliance on intrinsic (inner) motivation. To be *one's self,* the student needs to be free from the seductive pressures of evaluative praise, pressures which come from outside the student.

You might ask, "If evaluative praise can be destructive, why do students and others seek it out?" Praise *may* make the student feel good—momentarily. However, it creates a dependence on others, the teacher in our case. The teacher becomes the source of approval. The student relies on the teacher to fill his praise needs. The student needs the teacher to establish his self-worth. Student addiction to teacher praise gets in the way of the teaching-learning process.

Let's make the point of destructive praise a personal one. Can you remember a time when someone praised something about you and did it to an excessive degree or did it in such a way it was embarassing to you? Perhaps it was done in front of people you really didn't want to hear the praise and see the effect on you.

Perhaps somebody commented at length on the way you were wearing your hair or somebody praised a bit excessively a food dish you prepared or a project you made. Do you remember the feeling of "Oh, I wish he'd stop" or "Enough already!" You actually got uncomfortable. Of course, like all of us you were initially pleased that someone took special notice of you or your work and acknowledged you for it in some way—in this case by praise. The feeling of discomfort that comes from inappropriate praise, excessive praise, ill-timed praise, etc., is felt by students in classrooms all across the land.

Teachers use praise and why not! The reasoning goes something like this. "It was used effectively on me; look how I turned out.

Everybody likes a little praise now and then. What a terrible world this would be without people praising one another for what we say or do! And one more thing . . . praise works! Students work harder when they are praised."

To take praise from the teacher's repertoire of "influencers" is a serious undertaking. A great void would be left for the teacher who wants to acknowledge the good work, the good effort that students make. The void would be there if something were not being offered in its place. Something is available and can be used very effectively. That something is encouragement. We'll get to encouragement as soon as we conclude the indictment against praise.

We've been making the case for the fact that people (students) become uncomfortable with praise. We can even conclude that sometimes students get downright defensive whenever evaluative praise is used.

By evaluative praise we mean any statement that makes a positive judgment of a person, an object, an act, or an event that includes little or no supplementary information. For example:

"Good work, John, you've done a fine job."

"You're a very good student."

"That report of yours is excellent."

In the above examples you can see the judgment words, *good, fine, very good,* and *excellent.* You can readily determine that there is little supplementary information in the remarks.

Sometime, watch people whenever words like those used in our examples are used on them. You're likely to hear . . .

"Uh, well, thanks, but I had some help from ________."

"You're just saying that."

"Well I was just lucky, I guess."

"Oh, but look here. See the mistake I made here. . . ."

These are defensive reactions to an uncomfortable situation that praise has evoked.

Why do people react to praise with discomfort, even defensiveness? Part of the answer lies in the message that is inherent in praise. The message includes judgment of one person by another. Who likes to be judged! It also includes manipulation by the praiser.

That's right! When *you* praised someone for something, you wanted your praise to have a *certain effect* on the "praisee." You tell your friend Mark, in a genuine way, that the suit he has on is really magnificent. You made an appraisal of the suit and its suitability to Mark and you rendered your judgment about him. The praiser becomes the appraiser. A pecking order is established. You set yourself up as the judge, a status higher than your friend. Better to say simply, "Like that suit, Mark." Suppose one of your future students would say to you, "You are doing a good job, Teacher. Your mastery of subject matter is excellent. I am proud of you. Keep up the good work." Note how that elevates the student to higher status.

Teachers should and do have higher status than students, but not for judgment purposes. The status difference is for the teacher's greater knowledge, skills, experience and position, which is as it should be.

One more note on the perils of praising. We already mentioned there is an element of manipulation. In the example of Mark's "magnificent" suit, you dubbed it magnificent, letting him know you approved of it. You were also letting him know you wanted him to continue wearing suits like that—those that fit *your* tastes. Even if you protest and say "Well, I was only telling him what's best for the occasion. *He* put on the suit, etc., etc." You were still telling him *you* wanted him to continue wearing the suit on occasions like that. That's using your influence to manipulate him to *your* way of thinking.

In the kindergarten classroom, you might say, "Samantha, the orange in your fingerpainting is gorgeous!" You probably made that remark to keep Samantha productive and painting, or you wanted her to quit using so much brown in her paintings. The point is YOU wanted to maintain Samantha's current behavior so you put out the manipulative praise. Chances are, especially in kindergarten where children are eager to please the teacher, you're going to get a lot of orange from now on—in everything! Ask any art teacher who has had the misfortune of praising a certain characteristic of a fledgling artist's work what the effect of such praise can be.

There's more to be said for the negative effects of praise, but we need to get out of the condemning mode and into more of a helping mode. In our closure on praise we need to put it in perspective with two final points.

Number one, praise is a two-edged sword. It can, and usually does, cut for the bad, but it can also cut for the good. We mentioned the effect of praising the down-and-out, deeply discouraged student at least until he can be weaned away from the dependency-inducing effects of praise.

On the bad edge of the sword is the adverse cutting effects on students' self-reliance, self-direction and self-control. There is also the counterproductive effects that come from one person sitting in judgment of another, as well as the aversive reaction of the student who feels he is being manipulated. No one likes that arrangement.

Number two, evaluative praise will undoubtedly be used by you. It is an exceedingly difficult habit to break. It has almost been inbred in us.

If you're going to praise, stay completely away from praising the person as in "You're a good girl, Margaret." Instead praise the act, the product and by all means the effort the student made. An example of each:

ACT: "The way you handled that tough assignment was outstanding, Sarah!"

PRODUCT: "By gosh, Jim, your bread box is really well done!"

EFFORT: "Angela, you really tried hard to do that backward somersault. What a try!"

Even those three examples leave room for improvement. They are, however, an OK effort on the part of teachers to get to a better practice. What is that better practice, you ask?

Encouragement Is the Answer.

Encouragement needs to be used in place of praise. We identified Objectives 3 and 4 as objectives that will, if achieved, provide the skill to make you an effective, encouraging teacher.

Paraphrasing another great human relations thinker and teacher, Rudolf Dreikurs, we see in his words the importance of encouragement.

Like a Plant Needs Water to Grow and Thrive, a Student Needs Encouragement to Grow, to Learn and to Thrive.

Ginott came close to what we need to use in the classroom in place of praise when he developed the concept of appreciative praise. Here, we'll use the term *encouragement* as the alternative to the time-worn (out) practice of praising. Look at the 43 examples shown on the next page.

I don't believe I've ever seen one done like that; it's really one of a kind
Just wait till we put that on the bulletin board. Won't that be something!
I bet your mom and dad would be proud to see the job you did on this.
I'ts a pleasure to teach you when you work like this.
It looks like it's going to be a thorough report.
How can I be so lucky to have you in my class!
You can tell you take your work seriously.
That's an interesting way of looking at it.
Congratulations! You only missed _____.
This makes me glad I became a teacher.
You really gave that a solid effort.
Let's use your idea instead of mine.
You really outdid yourself today.
You're on the right track now.
Now you've got the hang of it.
Why don't you show the class?
My goodness, how impressive!
That's quite an improvement.
Everyone's working so hard.
That's coming along nicely.
Now you've figured it out.
That's the right answer.
I appreciate your help.
You may it look easy.
How did you do that?
Thank you very much.
You've got it now.
Very interesting.
You're thinking.
What neat work.
Exactly right.
Very creative.
That's clever.
Much better.
Keep it up.
Nice going.
Solid try.
I like it!
Mm mmm!
Aahh!
Ooo!
Oh!

Encouragement is at once two things. It is a social-emotional atmosphere in the learning environment which communicates to all who live, work and play in that environment that they are accepted as they are. It is also an active behavior on the part of the teacher which communicates to the student, "You can do it." We'll study these components one at a time.

To learn is to take risks, risks that are primarily tied in with the potential to make mistakes and to fail. Mistake-making and failing can be hazardous to one's mental and emotional health, hence the risk element. Elaborating on the classroom atmosphere component of encouragement, we know that several conditions must exist if the student is going to feel encouraged to take risks.

The teacher is in charge of developing the encouraging atmosphere. Once it's established, then it is the teacher again who must take care of the environment. In the truest sense of the expression, the teacher is the care-taker! That thought leads us to the first element that the skilled teacher will provide.

Components of an Encouraging Classroom Atmosphere

SIGNIFICANCE

Everyone in the classroom must have a sense of "I am cared about. Someone cares about me now and he cares about me and my future." In effect the teacher must generate feelings of significance among students. Each student must feel a sense of worthiness that comes from being a cooperating, producing member of the class. A sense of worthiness will inturn generate a feeling of significance in the group. It's much like living in the atmosphere of an "encouraging family."

RECOGNITION

There must be reliable opportunity to receive recognition for the student's progress and certainly for the student's effort to succeed. The best recognition comes not in the form of praise, of course, but in the form of appreciation by others, of joy and excitement, of others seeing the student succeed. Whether the recognition takes the concrete form of a student's successful work being displayed somewhere in the classroom or the verbal affirmation of fellow students and teacher for what was done, the recognition is vital to the student.

REALISTIC CHANCE TO SUCCEED

Every classroom must have in it something through which, with a little effort, every student will experience success. Teachers know that success begets success. Sometimes a particular learning task must be broken down into more manageable increments for some students. For others the task must be made even more challenging than what was planned. Regardless of the range of abilities in a given classroom, a skilled teacher must individualize and personalize her lessons so that every child, every day feels there is something he can do to experience success in the classroom.

ONGOING ACHIEVEMENT

One of the strongest motivators to any student is the sense of achievement. Knowing that he is achieving something is inwardly motivating and this motivating force is a powerful ally of the skillful teacher. Teachers must establish a system of ongoing assessment to bring out the fact that achievement is indeed going on. Sometimes the assessment is in the form of paper and pencil tests, sometimes in the form of a "Show-me-how you do such and such, Anthony." In the latter, the student shows the teacher and himself that he can do the task. Then the student feels the sense of accomplishment. Too often teachers think only that assessment is done to provide a basis for a grade or to satisfy school policy. Sometimes it is done to have grounds to support a case that the student is not doing well and the teacher knows she'll have to have "evidence" to support her case. Instead of such uses, assessment must be seen as necessary to give the student evidence that he is succeeding.

More can be said on the creation of an encouraging atmosphere, yet I'll close on this note . . . the teacher is the number one ingredient, the most important ingredient in that atmosphere. If the attitude and the behavior of the *teacher* is not in harmony with an atmosphere that encourages, encouragement will be a very, very unlikely product of human interaction in that classroom.

Speaking of teacher attitude, that brings us to the second of the two things that provide encouragement to the student. That "thing" is indeed the teacher, precisely the teacher's attitude and the teacher's behavior toward students.

The teacher's attitude, and behavior toward students, much like the components of an encouraging atmosphere, must communicate important messages. Let's look at four of them.

Behaviors of an Encouraging Teacher

UNCONDITIONAL ACCEPTANCE

By word and action, the teacher must communicate to each student that he unconditionally accepts the student as he is (warts, body odor and all). Please note unconditional acceptance is not the same as unconditional liking. There are no teachers who like all students! There are only teachers who, as human beings, dislike some students and especially some students' bad behavior. Whoever perpetrated the misguided notion on our profession that all teachers must love all students did us an injustice. Such a notion is not human and it gives teachers a chance to feel guilty (among other things).

Again, accepting is different from liking. Students have a right to be accepted as they are. Being liked is not a right. We have this idea that to be accepted you have to be "OK" first. The wisdom of a current bumper sticker is appropriate here.

You're not OK; I'm not OK.
But that's OK.

Students must feel that their teacher genuinely accepts them as unique individuals in their own right. Once students sense that the teacher accepts them as they are, then and only then, will they open themselves to the teacher's influence and be ready to change and grow and improve themselves.

EVERY STUDENT A WINNER

Teachers must find that special something present in each student, affirm it and reaffirm it as often as it takes so that every student comes to the conclusion, "I'm somebody!" By word and deed, the teacher must write on every student's psychological T-shirt a paraphrased version of the current message seen in education circles today. "God made you and God don't make no junk! You're a winner, kid!" Long message, but worth inscribing.

When students feel like they're winners and are worth something, they will learn what needs to be learned. No student will do well when he has the hangdog look of being whipped, of being a loser. In the lives of some students the teacher may well be the only person encouraging enough to give students reason to believe they are winners.

NONACCEPTANCE OF MEDIOCRITY

Never, never accept work that is mediocre, mediocre in terms of what the student realistically is capable of giving. To do so is to show a lack of respect for the student. By accepting shoddy work you are saying, "That's OK, that's all you're capable of doing." That message is a discourager. It is unrealistic to expect a student's very best effort on every task. That is not the point. The point is that you must reject work that is clearly below the student's personal standards. In refusing the work, you can say *in an encouraging way*, "You and I both know you aren't satisfied with this."

CONFIDENCE

It's the teacher who must send a constant message, primarily through action, that she believes the student can do whatever is realistically assigned. The teacher must believe CAN, even when the student believes CAN'T. How many times have *we* ended a successful venture with words like, "Geez, I didn't know I could do it!" It was very, very reassuring to know that someone believed in you, while others (and even you) didn't believe in your ability to do what you did. Students who have teachers who believe in students before they see successful performance are students truly blessed. That idea is a fit concluding statement for the confidence characteristic of the encouraging teacher. TEACHERS MUST BELIEVE BEFORE THEY SEE!

Summary

The purpose of this chapter was to get you thinking and believing that praise is not all that it is purported to be.

Praise has been around a long time and it has been a relatively unchallenged aspect of human interaction. Supposedly it always bodes well for the individual receiving the praise.

As an example of the lofty place praise has in our thinking, we can think of that eminent writer in the field of successful people,

Dale Carnegie. He exhorts us with, "Be hearty in your approbation and lavish in your praise." That idea is seriously at odds with the teacher whose mission it is to help students grow in self-reliance, self-direction and self-control. This chapter was written to shed light on the negative aspects of praise. If the case was made strongly enough, you have made a commitment to avoid or at least cut down on your use of praise.

The latter two objectives of the chapter were to have you understand the absolute value inherent in the encouraging classroom environment and in the encouraging teacher. As components of the encouraging classroom we analyzed the following four: significance, recognition, realistic chance to succeed, ongoing assessment. We also analyzed four behaviors of the encouraging teacher: unconditional acceptance, every student a winner, nonacceptance of mediocrity and confidence.

Through anecdote, exhortation, reference to experts in the field you may have become (more) committed to being an encouraging teacher. When you believe in encouragement, when you act on your belief in encouragement and when you actually encourage your students in ways of which you *are* capable, you become the best teacher you can be. And by the way, "God made you the teacher you are and God didn't make no junk!" (I just read that on YOUR T-shirt.)

Chapter 8

ENTHUSIASM

No commentary on effective teaching would be complete unless teacher enthusiasm was included in it. Let's use your experiences with your former teachers as conclusive evidence that teacher enthusiasm is vital to effective teaching.

Think back over all the teachers you can remember. Get a clear picture in mind of as many as you can. Now eliminate the ineffective or "bad" ones. Concentrate on the effective ones.

As you think about your effective teachers, ask yourself what made them effective. You answer, "Well, they knew and liked their subject matter; they liked kids (and they especially made me feel like they liked me); they made the subject seem interesting; they weren't bor . . ."

Hold it! Did you say "They made their subject seem interesting?" You did? Well, then, you just described one important aspect of enthusiasm. Whenever teachers can generate interest in the lesson at hand just because of the *way* they do something, they are teaching with enthusiasm.

Enthusiasm affects everyone. It not only *affects* everyone, it also *infects* everyone. Enthusiasm can indeed become contagious. As teachers we do want to affect and infect our students—in positive ways, of course. Before we get too far into this topic, we need to see what you are to accomplish in the reading of this chapter.

Chapter Objectives

Understand why students relate positively to the enthusiasm of the teacher.

Prepare a plan for being enthusiastic for the subject matter being taught and the teaching skills being used.

Prior to listing the two objectives, we were already making an effort to master the first one. Why is it students react and respond positively to a teacher's enthusiasm? One could easily respond to the question with the typical answer, "Well, no one likes a boring, unenthusiastic teacher!" That's true, but simply saying what we do not like does not say what we do like. More specifically, that answer does not say what will lead us to react and respond positively to the enthusiastic teacher. As professional educators, we can do better than the typical answer.

Remembering what we learned from the field of psychology, we realize that the human being has an inborn trait of curiosity. Curiosity is such a strong force within each person, it can be considered a natural striving. Curiosity is an ever ready ally of the knowledgeable teacher. A teacher can call on that ally time and time again. This was discussed in the use of motivation techniques and set induction strategies in Chapters 2, 4, and 5.

AFFECT

Curiosity fits into the question of how the teacher's enthusiasm affects the learner in the following way. A teacher begins emoting enthusiasm for a point of the lesson that was just begun. The student is curious to know, "What's she getting so excited about?" or "What's so special about that point that it has the teacher getting excited?" Enthusiasm can and will affect students positively due to their curious nature.

Students simply want to know, "What's all the fuss about?" That's why the question, "What's all the fuss about?" was invented—to satisfy the curiosity of those who didn't know and wanted to know!

CREATE A FUSS!!

Making a fuss is being enthusiastic (of course, our kind of fuss is a productive one).

INFECT

To show the infectious results of enthusiasm, we can go back to a point made in Chapter 2 on motivation. The point we made was that human beings need to identify with other human beings. There is a force inside each of us which impels us to imitate behaviors exhibited by others. It is a modeling process.

Whenever our students see us enthusiastically dealing with a learning task, they are quite naturally attracted to the task. They are attracted to it because they see us attracted to it. Because they identify with us, they will model our behavior. In this sense they are infected by what we do.

There are certain teachers with whom certain students will not identify. The influence of enthusiasm will not be so strong in such cases. Nevertheless the influence will be there because of the ever present effects of curiosity plus one more factor, commitment.

COMMITMENT

The enthusiastic teacher is one who exhibits an unmistakable commitment to what she is teaching. Whenever students sense the commitment there is a rubbing-off effect.

Who wants to be taught something when he feels the teacher doesn't believe in it? Teachers who want to impress upon their students the importance of the lesson at hand can do it in no better way than teaching it with gusto! With passion! With inspiration! With imagination! With energy! With excitement! With joy! With fun! With animation! With energy! Put those words in action in the classroom, one at a time, in combinations, or all at once if you can do it, and what do you have?

Enthusiasm!!!

The word *enthusiasm* is derived from Greek words which mean possessed by a god. An enthusiastic teacher indeed may be possessed by a powerful force, a force that can be called up at will.

On that last idea of calling upon one's own enthusiasm at will, you can be sure that any experienced teacher reading this chapter will agree on one thing for certain. There are times when teachers get ready to teach and they just can't seem to muster up the enthusiasm for the job at hand, but they begin teaching anyway. Six or seven minutes into the lesson they listen to themselves talking and watch themselves moving about and what do you know, they are surprisingly enthusiastic! How'd it happen?

Probably it was triggered by a student who became interested enough about the lesson to answer surprisingly well a question the teacher asked. Better yet, one of the students may have asked a good question. Perhaps the student was enthusiastic and communicated it to the teacher, who became "infected" by it. Teachers can be, and often are, influenced by their students!

The infectious relationship between teacher and student is often a reciprocal one when it comes to enthusiasm. It is usually a case of the enthusiastic teacher influencing the student to be enthusiastic, but the reverse cannot be overlooked. This reciprocal effect is closely tied in with the motivation factor of reciprocity we analyzed in Chapter 2.

There are some teachers who will say that whenever they work up an enthusiasm or whenever they get excited about something, it's a sure "turn off" for students. One of two things is at work here.

Number one, the lack of influence of the teacher's enthusiasm on students may be due in part to a lack of rapport between students and teacher. If the relationship is not a positive one, one in which the students *know* and *feel* they can be open, honest, frank, etc. with their teacher, the teacher's chance for influencing students is limited. Rapport building is called for. That building process is called for not only to get students receptive to the enthusiasm of the teacher, but also to other critical aspects of teaching. We won't go into those in this book, however.

A second reason that students may not be open to the teacher's enthusiasm is the probability that the teacher was not *truly* enthusiastic in the first place. The teacher may have been forcing, even faking it. Students seem to have built-in sensors for detecting insincerity and faking.

The advice for developing the teaching skill of enthusiasm is simple. Allow yourself to become a student yourself of what you are teaching and how you are teaching it. *What* you are teaching deals with the knowledge and skills inherent in the subject matter. The *what* includes the teaching of habits, ideals and attitudes that go along with the use of the subject matter. You also need to be a student in the practice of teaching. *How* you teach subject matter is the stuff of this book, i.e., the critical skills found in effective teaching.

Whenever teachers become students in their own subjects and of their own "craft" it never fails that a concommitant enthusiasm emerges. It is a natural, genuine enthusiasm, the kind that influences students in positive ways.

In the preceding paragraphs it was intimated that there are two kinds of enthusiasm. Indeed there are two basic types that we can identify and use for our purposes. There is an enthusiasm for one's subject and there is an enthusiasm for one's teaching. Let's look at the two separately.

ENTHUSIASM FOR ONE'S SUBJECT MATTER

Your students must see you doing things with your subject matter that sends out an unmistakable message, "This stuff is great!"

If you are . . .	do your students observe you . . .
1. an English teacher	reading or writing on your prep time and enjoying it?
2. an industrial arts teacher	turning a spindle on the lathe, whistling all the while?
3. a history teacher	checking into the local college library to do some research of your own?
4. a fifth grade teacher	doing some charitable work at the animal rescue shelter, which shows your commitment to the lessons you taught on kindness to animals?
5. a home economics teacher	making a surprise project for the field hockey team that includes a number of your students?
6. a business education teacher	taking dictation from the high school principal just so you can keep your skills sharp?
7. an art or music teacher	making a composition of your own and obviously enjoying it?

(continued)

If you are . . .	do your students observe you . . .
8. a reading teacher	reading a book when it's not a requirement?
9. a science teacher	bending over a microscope oo-ing and ah-ing as you see new things?
10. any teacher	moving about the school as a learner, appearing in the library, in the physical education rooms, in the computer room, etc.?

You get the point, I'm sure. It is exceedingly effective to have teachers displaying an enthusiasm for their subject matter. It communicates the idea that the teacher really does mean it when he says, "This subject I'm asking you to learn is really worth learning!" It's like the old saying some leaders use, "I wouldn't ask you to do anything I wouldn't do myself." There is merit to that idea.

Too often we hear teachers making comments like, "I teach this stuff all day. I sure don't want to do more of it on my free time!" Makes us kind of wonder if a commitment to the subject matter is really there in the first place.

Another "too often" we observe is teachers telling, telling, telling their students how important their subject matter is. They almost make their students "teacher deaf." Yet these same teachers who do all the telling, seldom do any showing of how important their subject matter is. When it comes to paying heed to what their teacher says or what their teacher does, which do you think the students believe? Do they believe the verbal or the nonverbal message? (Please underline the correct answer. You sure won't need a reinforcement cue for your response to that task.)

One final note on the need for a teacher's enthusiasm for her subject matter. Recall certain teachers' answers to questions students ask like:

"Why do we have to study ____ anyway?"
OR
"Why do we need this?"
OR
"We'll never use this junk anyway!"

Do you remember your teachers' responses?

"You'll need this when you get into college."
OR
"If you don't learn this subject you won't be able to learn the next subject."
OR
"You need it to graduate (be promoted to the next grade, etc.)"
OR
"Because it's required!"
OR
"Because I say you need it!"

Weren't those gems! (Perhaps you've even used them or something like them once or twice yourself). It is indeed pathetic whenever we cannot come up with valid reasons for why a subject is important in the lives of human beings. If we can't, then the one lament stands: "Why do we have to do this junk anyway!!!" The enthusiastic teacher will have a reservoir filled with reasons why her subject matter is important!

ENTHUSIASM FOR ONE'S TEACHING

For this type we will look at some research to see how enthusiasm was used for its positive effects on students.

Rosenshine and Furst[§§] placed enthusiasm third on their list of behaviors related to student achievement. The definition of teacher enthusiasm that was extracted from their studies showed the enthusiastic teacher as one who conveyed a sense of commitment, excitement and involvement with the subject matter. On the surface, this finding appears more supportive of the first type of enthusiasm we studied, for subject matter, yet it is also appropriate here. The teachers showed their enthusiasm while in the act of teaching. The researchers found that enthusiastic teachers made their lessons imaginative and stimulating, just like the teachers you chose as your most effective.

Gillett[§§] conducted a study in which he prepared one group of teachers to be enthusiastic in their presentations. Before the experiment the students were observed and diagnosed as being on task 75% of the time. That percentage is consistent with the time on task findings in other research. After receiving skill development

[§§]See Bibliography section.

in enthusiasm, the teachers then taught their students with newly mastered enthusiasm. Students' time on task went up to 86% task!

The message is clear. Teachers who increase their enthusiasm level will have a very high probability of positively affecting students' work level, that is, their time on task. More time on task directly correlates with increased achievement.

We'll look at one more research study. This one gets at the specific ways teachers can increase their enthusiasm. Collins[§§] developed an operational definition for enthusiasm. Her definition included some of the findings of other researchers. She identified eight indicators of high teacher enthusiasm. They are as follows:

(1) rapid, uplifting, varied vocal delivery
(2) dancing, wide-open eyes
(3) frequent, demonstrative gestures
(4) varied, dramatic body movements
(5) varied, emotive facial expressions
(6) selection of varied words, especially adjectives
(7) ready, animated acceptance of ideas and feelings
(8) exuberant overall energy level

As you contemplate each of the eight indicators and relate them to your own teaching enthusiasm, you will undoubtedly decide that you can improve in most of the areas mentioned. The question is, WILL you work to improve yourself? Committed professional that you are, there are no doubts. You will go to work!

Do you suppose you could be silly enough to stand in front of a mirror and work on indicator number two, for example—the dancing, wide-open eyes? Go into a room with a big mirror in it. (Lock the door behind you.) Look in the mirror and make like Eddie Cantor! Too old a reference? OK, then go for the Carol Channing look! Still too old? Let's see . . . Can you think of any current personality who has a renowned, enthusiastic set of eyes? When you do, make eyes like him or her.

Why not go over the eight indicators and demonstrate two examples of each. Do that and you'll be on your way toward making a personal plan for upgrading your enthusiasm. Sometimes you have to act <u>before</u> you plan. Doing so has a way of generating interest and commitment to ACT. By acting you'll likely do whatever you did again and again until you get it down the way you want it. Too often we teachers plan and plan and plan but we don't get the plan off the drawing board. We usually look for one more piece of information, or we look for the right time to implement the plan, or we just procrastinate.

The point is we act too late or we fail to act at all. Sometimes we should forget Ready, Aim, Fire. Sometimes we should Ready, Fire, Aim!!

Chapter 9

STIMULUS VARIATION[I]

Come with me to a U.S. Army base in Fort Knox, Kentucky. It is 1955. We're sitting in a room with approximately 50 other "students." Like you and me, they are recruits being trained to serve in the infantry. You've finished your 5:15 A.M. breakfast of powdered eggs, powdered potatoes, canned beef (circa 1943), powdered milk and real toast. It's now 6:15 A.M. and you've been ordered to pay attention to the upcoming 45-minute lecture on military transportation vehicles. The instructor's cunning is now coming to the fore.

"Because I know you're gonna nod off durin' my little talk on jeeps, trucks 'n stuff, I'm gonna arm this little 'splosive device. It has a timer on it so I'll set it to go off sometime during my talk. It won't hurt you but it'll probably scare the hell outtaya. But if you wanna go to sleep, be my guest." He puts the device on the front right corner of his desk.

There's no way you or I, or the other 48, are going to go to sleep, regardless of the boring nature of the lecture. Never mind the fact that our minds and emotions are set to go off whenever the " 'splosive device" goes off. Never mind the fact that the instructor doomed himself to an ineffective lecture because of an errant, invalid set induction. We're simply waiting for THE event. All that matters to us is THE event.

Somewhere about 25 minutes into the lesson there is a piercing BANG! It reverberates off the four walls, ceiling, floor and 106 ear drums (in case you're doing the math, that's two drums for the in-

[I]Stimulus variation as a teaching skill was originally developed for use in a teacher training setting by Dr. D. C. Berliner for the Stanford Center for Research and Development in Teaching, Stanford University, Stanford, California.

structor, 100 for our fellow infantrymen and two each for you and me). Moments after the bang, our nostrils (106 of those) pick up the pungent odor of burnt powder.

We've just experienced a facet of stimulus variation. The idea that the BANG was a stimulus won't be disputed, neither will the idea that an explosive device going off in a "classroom" is a variation on what is typical. This example serves as a beginning for understanding the teaching skill called stimulus variation.

The definition of stimulus variation is inherent in its name. It is the *variety* of events and experiences that occur in the setting in which the student is expected to learn.

The purpose of stimulus variation is to sustain the attention span of the student. In addition, it is used to rejuvenate attention whenever it is lagging or to reactivate it whenever it has been broken. It is a skill mostly used after a lesson is underway. Let's put this skill in perspective.

The instructional skills being set forth in this book are the means for implementing the planned lesson. The teacher plans the lesson; the teacher uses effective teaching skills to put that plan into action. Once the lesson's objectives or tasks have been made clear and once the motivation for learning and the set induction have been established, the lesson is underway. It will need a teacher to "tend to it" as it unfolds. Stimulus variation, along with reinforcement cueing, encouragement and enthusiasm, is a skill needed by the teacher to keep students learning.

Before going further, we need to understand that there are certain events or experiences over which the teacher has little or no control. While such stimuli affect the student, usually the effect is felt in unproductive ways. Examples would be a knock at the classroom door, an unexpected announcement over the P.A. system, a walk-in visit of the teacher's supervisor (or someone else), a piece of plaster falling from the ceiling (heaven forbid) and the first snowfall (heaven approve). These stimuli interrupt the lesson flow and even though the creative teacher can oftentimes use them to his advantage, they usually detract from what the teacher has planned.

For our understanding of stimulus variation we will consider only the stimuli over which the teacher has control. Some of these stimuli are as follows:

- bulletin board
- textbooks

- projection screen (for films, slides, overhead projections, filmstrips, etc.)
- maps, charts, posters, etc.
- the teacher
- things which students can taste
- guest speaker
- other students
- resource books
- recordings
- field trips
- video tapes
- models, mock-ups
- machinery, paraphernalia
- raw materials
- computers
- things which students can smell

You certainly can expand on the list, but you get the idea. As you looked over the list, you probably noted that all five senses are to be involved. We will see later that involving our five senses is an important consideration.

The list of stimuli represents what the teacher controls and can use to his purpose. Think back to the set induction of this chapter and specifically the explosive device set off by our cunning instructor. Was the device under his control? Yes, because *he* brought it to the classroom in the first place. *He* set it to go off at a certain time and in a certain place. *He* decided to use it as a stimulus to keep us alert. Of course, he unwittingly induced us to rivet our attention and our emotions more on it than on his sterling lecture, but that's another story.

There was one stimulus on the list to which you need pay extra attention. It is the most important of all. It is the teacher. It is YOU.

Yes, among other things that you are (or will be), you are a stimulus. Furthermore, you are a stimulus that must have variation. Otherwise, you can't be a stimulus variant. No self-respecting teacher would want to be less than a stimulus variant!

All levity aside, the fact that the teacher is the main stimulus in the student's learning environment is at the heart of this chapter.

The Teacher Is the Students' Primary Stimulus. As Such the Teacher Must Be Sure to Present Himself in a Variety of Ways.

The variation of the teacher as main stimulus can be done in at least five ways. Those ways represent the criteria for measuring effective stimulus variation.

MOVEMENT

The classroom or teaching station is usually rectangular in shape. The chairs in the room face toward a designed front of that rectangle. Typically, the teacher works at the front of the room. In the natural action of the teacher, she will move about that rectangle in varying amounts. The skilled teacher will be sensitive to the need to control those movements.

The effective teacher will time her movements, neither being in one place too briefly nor too long. She will be in the right place at the right time. She will be close to a student who is in need of the teacher close at hand, farther away from a student who is in need of distance from the teacher. She will appear in different places in the classroom: front, right side, left side, rear of room, center aisle, third aisle, etc. She will not teach with the windows as her background. Students would have to look into a bright background which can be hard on the eyes. Moreover, the background would have her competing with the birds, the trees, the clouds and other stimuli.

The teacher who appears in different parts of the room changes the direction of his students' primary stimulus. We know that the teacher who stands or sits at a podium and works from there for twenty to forty-five minutes at a time will have students fidgeting in their chairs. Fidgeting doesn't take long, perhaps ten to twenty minutes depending upon the student's developmental age. It simply is not natural for students to sit upright in a straight-back chair without seeking relief.

When the teacher moves to a new vantage point in the classroom, students' eyes and ears move to accommodate that change. When the eyes and the ears make an adjustment, the head likewise adjusts. When the head adjusts, so does the rest of the body. The physiological adjustments made by the body sustain and rejuvenate students' attention span.

Let's take an example. About once in a 42-minute period, the teacher appears at the rear of the room where a certain item is hanging on the wall. He makes sure his lesson plan includes a point that the wall-hanging will be used to support. He walks to the rear of the room while saying, "Students, if you will look at this drawing of Richard the Lion Hearted, you will see. . . ." In order for the students to look, they will of course have to alter greatly the way they are sitting. They will shift from sitting on two anatomical hemispheres to sitting on one. There will be an alteration in the flex of the spinal column. Will *you*, the reader, turn in your chair, face the opposite way, right now, so that you feel the effect your students will feel.

Didn't that turn do something for you? The teacher, as stimulus variant, can positively affect students' attention span by eliciting physical changes in their students.

A few cautions are in order. Teacher movement must not be taken to mean that the teacher is to be on-the-move constantly. It means that the teacher must move on occasion to different parts of the room and teach from there. A teacher who constantly moves about is a distraction, even an annoyance. Imagine what it would be like to be listening and watching a teacher who paces like the stereotypical expectant father. After four or five minutes of watching, you would come down with a severe case of the "dizzies."

Another caution is that the teacher must not teach from locations in the classroom which put some students at a disadvantage. For example, teaching from the rear of the classroom too frequently or for too extended a period of time, puts the teacher out of sight. Students will turn away from the teacher to write something in their notebooks or simply to relieve the strain of an unnatural sitting position. The teacher does not want to teach to the backs of students' heads.

A good rule of thumb is to appear at the rear once a class period. Of course where the teacher appears in the classroom is greatly determined by the arrangement of the furniture, the size of the class, the nature of the lesson, etc.

GESTURES

We touched on the benefits of gesturing as nonverbal communication whenever we analyzed the skills of reinforcement cueing, encouragement and enthusiasm. We did it in a general sort of way. Here we will be more specific.

The entire body can be used as a source for variation of the teacher as stimulus. Literally speaking, from the top of the head to the tip of the toe, the teacher can vary what the student is perceiving in her.

As an example, the teacher puts her hand on the top of her head and asks the class, "How many pounds per square inch is the air pressure on my head?" The question is more effectively put with the added variant of the hand-on-head gesture.

Another teacher says as a way of playfully making a point about people who talk too much, "He who thinks by the inch, talks by the yard, ought to get moved by the foot." In each reference to measurement the teacher uses gestures: her finger and thumb to approximate an inch, her two hands with arms extended to the front to approximate a yard, her right foot with toe flexed à la football placekicker Lou Grozza to show the action of the "foot." The teacher's point is more emphatically made and much more likely to be remembered because students now have a visual reference to add to what they heard. Gestures enhance the message.

Can you conjure up a picture of a teacher who never uses his face, his limbs or his body to enhance what he says? Unfortunately, you probably have had a teacher like that so it is not so much a matter of imagining or conjuring up an image as it is of recalling one. At any rate, the unanimated teacher can be deadly dull! A teacher does not have to have the gesturing ability of a mime, but it is imperative for the teacher who wants to be more effective to gesture whenever appropriate.

Let's see you sit still in your chair and not move *any* part of your body and teach to a make-believe student the meaning of the term, *spiral staircase.* Your make-believe student understands English and he knows what stairs are, but does not know what *spiral* means. Let's see you teach him without the use of any gestures at all. Oh, by the way, in this task you're not allowed to use any type of writing instrument.

Tough job, sans gestures! Fortunately, there is no law prohibiting teacher gestures. They are permissible, even welcome! Use them all you can, being mindful that even a good thing can be overdone.

By way of closure on the criterion of gestures as a stimulus variation, let's look at a modest listing of gestures in three basic areas: face/head, arm/leg, body.

FACE/HEAD: Eyes wide open, eyes squinting, eyes rolling, one eye winking, nose twitching, nostrils flared, lips pursed, lips pulled to one side of mouth, chin jutted forward, head cocked to side, head thrown back, head nodding, head motioning a direction.

ARM/LEG: Fist clenched for emphasis, right fist pounded into left palm for emphasis, fingers interlaced with arms close to chest showing reverence, hand stroking chin to simulate thought, hands outlining a shape in midair, arms folded, hand and arm overhead waving, arms behind back, legs spread shoulder width, legs spread extra wide, walking slowly/rapidly showing eagerness/anxiety.

BODY: Leaning forward (backward, left, right), crouched down, chest puffed out, chest pulled in, quivering in simulation of a chill or fright, standing on tiptoes, bent forward at waist, squatting down by bending at knees.

You can and you will add to the listings. They are by no means complete. You ought to use, freely and convincingly, those that come naturally to you. Those which do not come naturally, work on and incorporate into your teaching. Teaching with animation is teaching with greater effectiveness.

FOCUSING

You're standing fifty feet away from a long-awaited event. You are a photographer and you have been given the assignment of photographing a ruby-throated hummingbird. Specifically, you have been told to photograph it as it penetrates a rose-red flower with its long, slender, snorkel-like beak. You have thought of everything. You now have your camera with zoom lens at the ready. You deftly raise the camera with the viewing window to your right eye. You have the urge to get the best possible picture you can.

What's the first thing you would do with the camera? (We'll assume you have taken the lens cover off.)

Of course, you would want to make sure you have that bird in focus. That's the advice for this criterion of stimulus variation. "Get that bird in focus!"

There are numerous times in the course of a lesson when you have made, or a student has made, a significant point. You will be concerned that the point may be lost so you want to "zoom in" and focus on that point. How do you do it? "You get that bird in focus!"

You can do it by telling your students in some way you want them to focus on the point. You can also, by your gestures, cue them to pay attention to the point. Thirdly, you can use a combination of telling and gesturing.

Let's "zoom in" on some examples of the three possibilities.

Focusing by Words

(1) "Pay particular attention to the fourth reason, class."
(2) "Notice what happens when I release the latch."
(3) "Watch closely, very closely . . ."

(4) "Would you just look at that!"
(5) "Pick out the fifth word on the third line . . ."
(6) "Keep your eyes on the ball as it rolls down the plane."
(7) "Stare at this red book cover for one minute and . . ."
(8) "I can't believe what I'm seeing in the background of the picture on page 16!"

Focusing by Actions

(1) Holds up object with left hand and snaps her fingers behind the object to get everyone's attention to that spot.
(2) While the class is looking at a screen, masks off an overhead transparency so that only one sentence, phrase, or word shows.
(3) Uses hands as parentheses to enclose a key phrase or word written on the chalkboard (or projected on the screen). Puts hands into an arc-like parentheses and places them over the parentheses of the third example to get the effect.
(4) With knuckle or hard object, raps chalkboard where the key point is written.
(5) Uses pointer, ruler, pencil, or some other pointed object to indicate the key point.
(6) Underlines, draws ellipses or circles around the key point.
(7) Puts stars, asterisks, or some other eye-catching symbol in front of or behind the key point.
(8) Becomes more animated as the key point is made.

Focusing by Words and Actions

Take the eight examples of telling and the eight examples of gesturing and use one from each in any combination you wish. Doing so, you will have 64 examples for this criterion!

SHIFTING SENSORY CHANNELS

We know that the more senses a person uses when he or she is trying to learn something the more likely it will be learned—and remembered. Knowing this, we must commit to our repertoire of skills the practice of tapping into as many of the five senses of students as we can. That idea is so basic that it hardly seems worth mentioning. However, you can visit many classrooms and watch teachers relying almost exclusively on the students' sense of hearing. Hearing is followed by seeing as a distant second. Feeling is an even more distant third. Smelling and tasting? Well, they are as rare to the typical classroom as that ruby-throated hummingbird is to your backyard!

As an illustration of a teacher tapping student senses to enhance learning, I was privileged to observe a highly effective lesson taught by a fourth grade teacher. He had taken one of his three reading groups through several reading lessons that had to do with fishing. As a culminating activity he schooled his students in the Alpha and the Omega of hosting a fish fry.

The students learned how to buy fish, prepare fish, select the right side dishes to go with a fish entree, set the table, make the placemats for the table setting, write and send invitations, greet visitors as they enter the room, wait on the tables and clean up afterwards. They saw, they heard, they felt fish, they tasted fish, and what else? They "smelt" fish. The fish was haddock, however, not smelt. (Ouch!) None of those twenty-four students will forget their activities dealing with fish.

Of course, every lesson, every day cannot engage all five senses, but whenever we can engage three or four, we must do so.

The criterion of stimulus variation under analysis is "*shifting* sensory channels." We must concentrate on the first word, shifting. The idea is to shift from one channel or sense to another, not in random fashion, but with design and purpose.

As with teacher movements, the teacher must be conscious of the amount of time she spends in one area, in this case the sensory channel. She also needs to be mindful of the timing when a shift is called for. Finally, she needs to be skilled in the strategies of beaming to one channel rather than another.

More than one channel can be used simultaneously. For example, a student could see and smell H_2SO_4 while he listens to his chemistry teacher saying that sulphuric acid is a powerful chemical. Three senses are engaged!

When the teacher determines that a shift is needed, the shift should be made. The determination ought to be made in accord with the *students'* need for a change of channels, not the teacher's need. For example, the teacher notes that several students are showing frustration in their facial expressions. They don't understand the teacher's explanation, even though it might be a second explanation.

The teacher detects the nonverbal student message of near frustration on the hearing channel, so he *shifts* to the seeing channel. The teacher draws a diagram on the chalkboard. After a brief encounter with that, he realizes his seventh graders just aren't grasping the concept he is teaching. (The concept is surface tension of liquids.) The teacher remembers his days as a football quarterback and says to himself, "Ready! Shift! 1 - 2 - 3! Hup!" So a shift is made.

The feeling channel is tapped. This time the teacher puts a sewing needle on the palm of each of the four students' hands. He tells them to bounce the needle up and down in their palm, to get the *feel* of the weight of the needle. Having accomplished that, he invites each one in turn to come to the front of the room and carefully drop the needle on the water surface of a glass of water.

"Be careful now," the teacher says. "I want you to drop the needle from a distance of about one quarter inch above the surface of the water. First make sure you are holding the needle lengthwise and parallel to the surface of the water. Do not let the needle drop point first."

The students do it and *voilà*! The needles float! Not one, not two, but three needles float. The fourth student "dive-bombed" his needle. What else! The teacher can use that fiasco to teach what happens when directions are not followed! Also the class can determine why that needle broke the surface tension and the other three did not.

The above example illustrates all three considerations for shifting sensory channels. The teacher spent a certain amount of time (perhaps too much) on the hearing channel; the *timing* of the shift was right because frustration was just beginning to show; the strategy of using a drawing was called for and when that didn't pan out,the teacher wisely shifted to a third channel. The channels used were hearing, seeing, and feeling/seeing. This is the way teachers must work if they are to be effective in sustaining and rejuvenating students' attention span.

PAUSING

Coca Cola had a saying that is apropos here, "The Pause That Refreshes." Pausing at the right time can indeed be refreshing to the students' attention span. That's what stimulus variation is all about . . . keeping the student fresh and refreshed and the teacher refreshing! Since pausing is a criterion of effective stimulus variation, its function is to sustain and rejuvenate or refresh students' attention span.

We begin the analysis of this skill by saying what is meant by pausing and what effect pausing has on the student. After those prerequisite understandings are achieved, we'll see how the effective teacher goes about the important business of pausing.

Pausing, simply put, means stopping, for a time, what you're doing or what you're saying. The pause could be a second or it could even be a minute or more. The kind of pause to be analyzed here is the kind that involves silence. Silence is an essential component of pausing. Pausing does not mean stopping what you're doing and moving to something else. That strategy involves a shift and not necessarily a silence.

The student finds the pause is necessary because often he needs time to get "the point" so that learning can "sink in," as the expressions go. Too often teachers make a profound point or a difficult point and they do not give the student enough time to internalize it!

Some students need more time than others. The length of the pause will vary with the student, but the fact remains that all students need "think time." Think time occurs whenever teachers incorporate the stimulus variant of pausing into their lesson presentation skills.

The pause also affects students in other ways. For example, teachers say they want their students to become good questioners. A primary question must immediately be raised: "How can we expect students to become good at questioning if we don't allow them time to contemplate a point being made and to think of a question to ask!" Students may need even more time to phrase the question properly.

Think about the times when you asked a question of a teacher or speaker and how you mulled the question over in your mind. You phrased it, rephrased it, thought you'd like to ask it, and then thought it might make you look foolish, then you thought you'd ask it anyway, etc., etc. All that took time. And what's the teacher or speaker typically doing while you are mulling? Going right on! And what about you? Well, you're missing much of what's being said or done because you're concentrating more on your question than on the teaching. It's tough to mull and attend to the teacher at the same time!

Effective teachers are sensitive to the mental processes of students. They provide time for processes to take place in an air free of competing stimuli, chief of which is the teacher's voice.

Another effect of the pause on students is the signal it sends. Students know that whenever there is a lull in the action something is up. Such is the case when the teacher suddenly stops talking. The student senses the disruption in the flow and attends to the stimulus variant (the pause). The student's internal commentary goes something like, "What's that? Oh, the teacher quit talking. Something must be going to change. Nope, nothing coming. Ah, then something must have just been said that I'd better know. What was it? Oh, yeh, 'The square root of the sum of the squares of the base and the height of a right triangle gives the length of the hypotenuse.' I better write that down. Besides, it'll look good if she sees me taking notes."

How long did the internal dialogue take? Probably three to four seconds. That's all. The length of the pause necessary for effective stimulus variation is not really that long. The pause is necessary, however, if we want students to think about what we say and do.

As indicated, there are numerous positive effects that pausing can have on students. A partial list is provided below. It begins with the points already made.

(1) Allowing a point made "to sink in."
(2) Forming a question the student would like to ask.
(3) Signaling the student that something important was just said or done.
(4) Isolating segments of a complex lesson into more manageable units.
(5) Forming an answer to a question the teacher has just asked (we'll look at this in greater detail in the next chapter).
(6) Feeling a surge of curiosity about what will be said or done next (suspense-building).
(7) Practicing listening skills.
(8) Resting from a steady or intense learning pace.

Pausing can also be used as a disciplinary strategy. Using the pause in elongated fashion, i.e., letting the misbehaver(s) feel the weight of silence can be very effective for correcting misbehavior. Such use of silence is outside the scope of this book, however.

Summary

In this chapter we have analyzed five components of the teaching skill stimulus variation: movement, gestures, focusing, shifting sensory channels and pausing. With each of the criteria we saw that a variation of what was being done at the time evoked a change in student behavior. That change brought about better student attention to the task at hand. Sometimes the student's attention had to be recaptured if it had been lost. Any or all of the five criteria can be used effectively to bring this about.

Stimulus variation, along with the teaching skills of reinforcement cueing, encouragement, enthusiasm, and questioning, are all seen as mid-lesson presentation skills. Stimulus variation is used after the lesson is begun and the teacher finds it necessary to provide some *variety* in the lesson to stimulate the student in the learning process.

By the way, did you notice that something was missing from this chapter? Think about it for a moment. . . .

There was no stated objective for you to accomplish by your reading-thinking process! If you are now thinking it didn't make much difference, please think again. Suppose you were asked by a master teacher, someone whose judgment and skills you really admired, what was important about stimulus variation. What would your answer be?

You might be surprised that your answer will vary, even greatly, from others who read and thought about the same material. While differences could be all right, there is a very good chance you would be missing something. That something could well be a critical part of the teacher's message. Specific objectives provide consistency in learning among students and they keep teachers and students more constantly *ON TASK!*

To make the point, please review the headings of this chapter and see what three (3) objectives you might write down. Perhaps your writing will give you a little practice on what you learned in Chapter Three, "The Objective." Your struggle to get the *right* three objectives will make the point that it is not good to have students guessing about what they're supposed to learn!

Chapter 10

QUESTIONING

The teaching skill of questioning is so comprehensive entire books have been written about it. This chapter will be a modest presentation of what the effective teacher must know about the questioning skill. With a diligent effort to assimilate and use the ideas put forth in this chapter, you will go a long way toward overcoming the "Three-Quarter Rule" which plagues our profession regarding the skill of questioning.

"What's the 'Three-Quarter Rule,' " you ask? Stay with the chapter and we'll get to it. It is a very important rule! Let's hope your curiosity and your motivation are aroused because that's the idea for getting you into this chapter. We will get to the Rule, but first things first.

As our set induction for this chapter, let's use the question itself to get into the topic of questioning. You are about to be asked seven (7) questions. Each will require an appropriate answer. If no one is within hearing range of you, please answer out loud. If there is someone nearby, you may handle your answers another way.

The Seven Questions

(1) What's the name of this book you are reading?

(2) Please look at the dominant color on the cover of this book. Do you recognize it?

(3) In your own words, tell why it is important for a teacher to have command of the set induction skill in order to be an effective teacher. (Note that number three is not put in the interrogative form as most questions are. For our purposes, it has the same effect as the interrogatively

worded question, therefore, we will refer to it and others like it as questions.)

(4) Think of an example of a positive *nonverbal* reinforcement cue that you will use in your teaching and demonstrate it to an imaginary class of students seated to your left.

(5) Which of the five criteria of stimulus variation is most closely related to the criterion of teacher nonverbal behavior in reinforcement cueing?

(6) What mnemonic device can you create so that you remember t_cCIR? (By mnemonic device, we mean some mental "gimmick" that will help you to remember the five letters. For example, a mnemonic device for remembering the colors of the spectrum is the name ROY G. BIV. It helps many of us remember that the colors of the spectrum are Red, Orange, Yellow, Green, Blue, Indigo, Violet. To remember t_cCIR, you might want to create a sentence like the cat's Claws Indeed Rip!).

(7) Which of the seven teaching skills covered in Chapters Two through Nine is most important to you?

Now that you have recited your answer to each of the seven questions, you may move from student to scholar. You're going to be asked to do something different with the same seven questions. Please go back over the questions, one at a time, and think about the *personal* mental process you used to "get your answer." Before you do, however, let's see how this second assignment of the scholar is different from the initial assignment of the student.

When you were being asked to answer some questions on material you have been studying, you were being tested on what you were to have learned. Such is the role of a student. Now you are being asked to return to the questions and do something more advanced than merely answering questions. Such a request enables you to function on an authoritative level. Frankly speaking, you *are* becoming more of an authority in the field of teaching because you are mastering a body of knowledge and skills that others do not have. You have reason to feel good about yourself!

Now that we have cast you in the mold of scholar, you are ready to go back to each of the seven questions and get in touch with the kind of mental activity it took to come up with the answers. Please begin your task. . . .

Back already?

What did you have to do mentally to arrive at your answer to the question, "What is the name of this book you are reading?" If you answered something like:

"I had to remember"
OR
"I had to recall'
OR
"I had to think back"

then you are on-the-money with your answer.

Let's see what you did with another question, a more difficult one this time. . . . Look at number 5. Here you were taking apart one skill, stimulus variation. You were looking at one of its five components in relationship to the nonverbal cueing component in the reinforcement cueing skill. Such a mental process is usually identified as *analyzing.*

As we develop this chapter together, we'll get at the other five answers and you will see how scholarly you really are! You will do well (a little encouragement there . . .) .

If your appetite is whetted to learn a bit more about questioning as a teaching skill, a skill that must be used by effective teachers, then it's time for our objectives.

Chapter Objectives

Understand that the use of valid and reliable questions is a critical part of effective teaching.

Know how to formulate questions that provoke different levels of thought.

Know how to phrase questions that are free from five common pitfalls or errors:

—spoon feeding
—ambiguity
—compound questions that are not intended to be so
—unintentional eliciting of yes or no answers
—"unquestions"

Now for the infamous "Three-Quarter Rule!" (Hope your curiosity was aroused!)

Walk into any American classroom at any time and in any subject or grade level. You will find that three-quarters of the time someone is talking. Three-quarters of that time it will be the teacher talking and three-quarters of *THAT* time the talk has to do with a QUESTION!

Any way you look at it, there is a whole lot of questioning going on. The three-quarters factors used in the rule are approximations, to be sure, but they are exceedingly close to what researchers have found. For example, the final three-quarters factor was actually a 40% to 80% finding by researchers. Teachers deal with questions 40% to 80% of the time they are talking! No matter the preciseness of the rule, the facts thunder the message that teachers use the question in staggering amounts!

Let's ask a BIG question here. How effectively do teachers typically use the question? The answer is—not too effectively!

Let's Change That

In the mid-1950s Benjamin Bloom and others developed an important system[§§] for classifying educational objectives. The system later became used by educators for more than classifying objectives. It became a valid way for classifying questions.

Bloom's *Taxonomy* provides the basis for classifying questions into different levels of thought. Levels of thought explicitly tell us that there are higher and lower levels of thinking in which students can function. Sometimes teachers want their students to think at higher levels, sometimes at lower levels. It will depend on the nature of the task to be learned, the type of activity the teacher plans for accomplishing the lesson objectives, the kind of test to be administered, etc. The point is the question must be used as an effective means to engage the student in varying levels of thinking.

Asking questions designed by using the *Taxonomy* can elicit thought in one of six basic levels. The levels in ascending order are:

(1) Knowledge (recognition and recall)
(2) Comprehension
(3) Application
(4) Analysis
(5) Synthesis
(6) Evaluation

As students move in thought from a lower level, say comprehending, to a higher level of thought, say evaluating, they are *usually* thinking of things that are more difficult to master. For example, a seventh grade student is asked, "How does the mosquito fly?" That question requires the student to understand the fundamentals of flight and something about the anatomy of the mosquito. Such understanding presents a moderate degree of difficulty. This is a Level 2 type thinking.

On the other hand, a seventh grade student is asked, "Analyze the anatomy of the bumblebee and explain the bee's ability to fly in spite of scientists' belief that the bee's anatomy makes it impossible to fly." Such an analysis presents a great degree of difficulty. This example illustrates Level 4 type thinking. Level 4 usually requires more cognitive effort than Level 2. This is true of other upper levels of thinking in contrast with lower levels.

There are times, however, when the lower level can require greater cognitive effort than the upper level. For example, a teacher asks the question, "What are the latitude and longitude designations for Bombay?" The question is meant to be mere recall (Level 1) for students who have been studying the geography of India. It could require great difficulty to recall such specific information, especially if the information had been introduced among numerous other geographic facts.

If those same students had been asked a Level 6 (Evaluation) question such as "Students, think about the climate and location of Bombay. Would you like to live there?" Here students only have to assess their own likes and dislikes regarding climate and location and decide whether they would like to live in Bombay—not too difficult a thinking assignment. In this case, a Level 6 question is easier to answer than a Level 1.

One further point about the *Taxonomy* is in order. Basically, the six levels are hierarchically arranged. Let's look at an example. In order for students to function at Level 3 (Application), they must first be able to function at Levels 1 and 2 (Knowledge and Understanding), with respect to the learning task at hand. A teacher might ask her fourth grade pupils how they would write a word she just pronounced for them. The word is *change.* The students are to apply (Level 3) what they have learned. To do this task, they must first know (Level 1) the six letters, *c, h, a, n, g,* and *e* and they must know the consonant blend *ch* and the rule that the long *a* sound of change will require a silent e on the end of the word.

They must also understand (Level 2) the sounds of *ch, a, n,* and *g,* and which letters are used to symbolize those sounds. What the pupils *know* and *comprehend* they can apply. The thinking skill of Application includes the skills of knowing and comprehending, which illustrates the hierarchy.

There are occasions when a student can function at a higher level without the benefit of mastering the lower levels of thought. Often a student will function at the creative thought level of synthesis (Level 5) without the need for knowing, comprehending, etc. Such activity is outside the scope of this chapter, however.

Now that we have a working understanding of the *Taxonomy*, we're ready to look at some questions that exemplify the six levels. You probably noticed that examples have been sprinkled into the preceding paragraphs. Ah, but did you notice *how* we asked the seven (7) questions as the set induction for this chapter? (You did! Son of a gun, your insight is truly unique!) Yes, those questions illustrate the six levels of the *Taxonomy.*

Let's go back to the seven questions and use them to exemplify and explain each of the six levels of the *Taxonomy.* Seven questions to explain six levels! Two into one must go . . . and they do.

Knowledge

The first two questions exemplify Level 1. You were asked:

"What's the name of this book you are reading?"
AND
"Please look at the dominant color on the cover of this book; do you recognize it?"

In the first question you had to *recall* or remember something that you may or may not have learned. In the second question you merely had to determine whether you *recognized* a certain color, usually a very easy thing to do. Recognition and recall of information are the types of thinking required for Level 1.

Examples

When is Columbus Day?

How many players are needed for a basketball team?

What is the action word in the sentence?

Who spun the web?

Define the word *caustic.*

Comprehension

The third question you were asked was, "In your own words, tell why it is important for a teacher to have command of the set induction skill in order to be an effective teacher." To answer that question you had to *know* certain concepts like connection and you had to grasp the *meaning* of the concepts that make up the set induction skill. You had to understand what you know. Some other action words or verbs that get at the heart of Level 2 Comprehension are: *explain, compare, contrast, interpret, translate,* and *extrapolate.*

Examples

Explain in your own words why Mr. Roberts says he's your friend.

Compare Mexico's and Canada's reasons for wanting friendly relations with the United States.

Contrast silk and satin in terms of their composition.

What is the interpretation of the line, "Water, water everywhere and not a drop to drink?"

Translate *arc en ciel.*

What do you think space travel will be like in the next century?

Application

You were asked, "Think of an example of a positive *nonverbal* reinforcement cue that you will use in teaching and demonstrate it to an imaginary class of students seated to your left." You were prompted to *use* what you had learned. Hence your cognitive process was one of applying what you know and understand. If you gave a nonverbal cue to your class like the thumbs up sign, your physical overt act was directed by Level 3 thought.

Examples

Class, use the word *Rotterdam* in a sentence.

What is the best location on the map for growing wheat?

If we allow Bobby to bring his gerbil to class to live with us what do we have to do beforehand?

Give an example of what you just said.

Look at the definition we just finished writing, then tell us which of the three terms best fits.

Analysis

For this level of cognition you were asked a fairly difficult question: "Which one of the five criteria of stimulus variation is most closely related to the criterion of teacher verbal behavior in reinforcement cueing?" To arrive at the best answer of *gesturing* you had to do some high powered analyzing. You had to sort out the five elements or criteria of stimulus variation and then sort out the elements of reinforcement cueing. You had to know, understand, and be able to apply certain things you learned about the criteria. You then had to think critically about these elements, finding their commonalities and differences. Finally, you had to decide on a best fit of one criterion each from stimulus variation and reinforcement cueing.

Basically, analysis-thinking requires the student to break down material into its component parts and understand each part separately as well as the organizational structure of the whole.

Examples

From our talks on prejudice in the workplace, what are the two most important things an employer should consider when hiring employees?

After looking at the two photographs, tell us what you think the girl might have done before she began crying.

Analyze the idea that if 18 year olds are old enough to fight for their country they're old enough to vote.

How did the artist use the colors of gray and black to express his sadness?

In what ways do the three branches of the government keep officials from getting too much power?

Synthesis

Did you develop a mnemonic gimmick for remembering the letters t_cCIR? If you did, you answered the synthesis question,

"What mnemonic device can you create so that you remember t_cCIR?" Your answer was a creative one. You generated something new and different. You put parts together to form a new whole. In a word, your thinking was divergent. Up to this point, the questions directed to you prompted convergent-type thinking, that is thinking that directed you to a "best answer."

Thinking at Level 5 (synthesis) opens thought to new possibilities. If you want your students to be creative thinkers, synthesis is the level of thought you want to elicit. Questions at this level often have the words, "What if?" "Suppose . . . ," "Write a new (ending) . . . ," "Design a different . . . ," "Create. . . ."

Examples

Write a poem to tell how you felt when your pet died.

What would be the best solutions for putting out fires in a house if there were no water available?

Make a plan for talking the mayor into visiting us.

Create a crossword puzzle that uses the six new words from today's spelling list.

Evaluation

The last of the seven questions asked for a judgment or preference on your part, "Which of the seven skills covered in Chapters 2 through 9 is most important to you?" Evaluative thinking is related to the ability to judge the value of material for a given purpose. The judgments are to be based on criteria. They can be of the student's choosing or they can be inherent to the organization of what is being judged, or they can be criteria that are supplied by others. Our examples should help us focus on evaluation-type thinking.

Examples

Based on what you are feeling now, tell whether you think the music is good or bad (student's own criterion).

Was Bobby right when he turned and ran from the boys who said they were going to hit him? (Teachers' or parents' advice on the best way to handle such situations will be the likely evaluative criteria used by the student.)

Check which ones are prettier in terms of color blends and smooth lines (external criteria supplied by someone else).

There you have the six classifications of Bloom's *Taxonomy.* They are the basics. Each classification has subclassifications which you may want to pursue. To that end book references will be listed in the Bibliography.

A final word of caution is in order on the forming of a question so that it elicits the desired level of thinking. Sometimes the teacher will form a question that is intended to elicit one level of thought, yet the student will respond from another level. A teacher might ask an analysis-type question, and the teacher doesn't know that some students have been asked the question before. Therefore, students answer the question from recall rather than analytic thought. That is neither good nor bad. That simply is the way it happens sometimes.

Here's an example of a question being formed at one level and the student answering at another:

TEACHER: Please use the word Rotterdam in a sentence.

STUDENT: My sister ate all my candy and I hope it'll rot 'er damn teeth out!

Our teacher was working on application and the student on synthesis (or perhaps in another domain—the affective).

We need to get on with the proper phrasing of questions. It is likely that the proper phrasing of a teacher-formed question will elicit both the right level of thought and the answer(s) sought after. Chapter objective number three deals with proper phrasing and the avoidance of certain questioning pitfalls. We'll list the pitfalls in order and explain them. There will be examples of wrong and right phrasing. Understanding what you read, you ought to accomplish the objective of knowing how to phrase questions effectively.

PIT #1: Spoon-Feeding

Here's a serious but common pit teachers fall into. Once they're in it they work hard to stay there! They don't realize how tired they make themselves.

You know you're in this pit whenever you hear your voice saying, "Geez, these kids don't want to do any thinking!" or "It seems like I'm doing all the thinking today!" or the ever popular, "It's like pulling teeth to get anything out of these kids!"

What's gone astray is that teachers, from the beginning—kindergarten right on through grade 12—spoon too much information into their questions. They heap the information on so high that when the student has the question finally put to him he has only to grunt out a monosyllabic reply. My favorite example of this doesn't come from the classroom and a teacher. Rather it comes from a lawyer who was helping a Board of Education interview candidates for a finance officer's job. Even though the example is not from the teaching activity per se, it still makes the point.

LAWYER: Have you ever had experience in dealing with sheriff's sales, and by that I mean certain properties will be put up for sale because the owners have failed to pay their taxes on the properties and they must be sold off to pay the lien . . . any experience with that?

CANDIDATE: Yeh.

We still don't know whether the candidate knows anything about sheriff's sales! The same is true when teachers spoon-feed students with questions like, "Boys and girls, the word in a sentence that modifies a verb, an adjective, or <u>another adverb</u> is a what?" This is not only spoon-feeding, but it is giving the answer to the observant student. Reread the question and note what the two underlined words tell the student.

Teachers spoon-feed all too often and for several reasons. They do it unconsciously; they do it because it's the only way they seem to be able to get students to give *any* answers; they do it because they want to help certain students give answers who wouldn't give them if they weren't spoon-fed. All of these are self-perpetuating problems which ultimately are self-defeating. If we want students to learn to think, we must teach them how. The effective question is vital to the process.

On a related matter, but off the topic of spoon-feeding for the moment, is the sin we commit by providing too brief a time between the asking of the question and the assigning of a student to answer it. Research has borne out a rather amazing statistic in this regard. The time teachers wait between asking and assigning a question is about one second! That is not a lot of time to think of an answer. If you extend the wait-time before assigning the question you will simultaneously extend the think-time and you will get better answers. And research tells us that students' answers themselves even get longer! Back to spoon-feeding.

Spoon-feeding is a pit teachers fall into because they sense that too long a delay in getting any answer, let alone a right answer, means something is wrong. They rush in to fill up dead air time with a rephrased question to help their students come up with an answer. Let's be quick to point out that sometimes a rephrased question is in order. There are times when the question asked was too complicated or the students didn't have the background information to answer it.

Now for some examples that lurk in PIT #1:

WRONG: The square root of the sum of the squares of the height and base of a right triangle is called an ____________.

RIGHT: Define *hypotenuse.*

WRONG: Ludwig von Beethoven, one of the world's greatest musical composers, had a physical problem which he overcame and it was what?

RIGHT: What serious physical handicap did Beethoven overcome?

You can think of many more examples. You will want to stay out of this pit, but like the rest of us imperfect teachers, you'll fall in every now and then. Just remember not to stay in it too long. It represents a real deterrent to student thinking.

PIT #2: Ambiguity

Teachers fall into #2 fairly often. They fall because they do not give adequate forethought to the key questions they intend to ask at crucial points in the lesson. Failing to do so requires the teacher to think on her feet for the phrasing of the question. Such thinking often results in questions that are unclear to the student.

Excessive words, ill-chosen words, stammers and stutters, all too often creep into the impromptu question. The result is confused students. Here's an example:

> "Why do you feel, er, that is, what do you think is the main reason rabbits are so numerous? By that I mean, why do rabbits have so many offspring—other rabbits that is?"

Here's one for the books:

> "I know that you believe you understand what you think I said, but I am not sure you realize that what you heard is what I meant, isn't that right?

Gives you a hernia just trying to figure it out!

We all ask rambling and confusing questions now and then. But we do it more often whenever we only have a skeleton of an idea of what we want to ask and we sort of "flesh it out" as we are talking out loud. We need to spend time and effort thinking through, and in precise language, what it is we want to ask in the way of the key questions.

The well-phrased question is so vital to effective teaching that no lesson plan should be conceived without its presence. At the very least, the typical six to eight key questions in a half-hour lesson should be thought out in advance of the teaching of the lesson. It's an even better idea to *write out* those questions in advance. Writing often prompts the teacher to even greater clarity.

Here are some questions in our WRONG-then-RIGHT format which will help make the point of ineffective teaching whenever the teacher asks vague questions.

WRONG: What about the ancient Greek's commitment to mind and body?

RIGHT: What was the ancient Greek's commitment to mind and body?

WRONG: Is the wording of my question too copious for your diminutive comprehension?

RIGHT: Is my question too difficult?

PIT #3: The Compound Question

Simply put, teachers must avoid the pitfall of asking compound questions whenever it is better to ask questions one at a time. There are times when it may be all right to ask two questions at once, but one of the questions ought to require quick thought and a brief answer, while the other might be a bit more demanding in thought and answer.

For example, you might ask the compound question, "Who made the first trans-Atlantic flight in an airplane and why did he make it?" You can see where the first answer is fairly easy to recall. It could even help stimulate the memory a bit to retrieve other cognitive information needed for the answer to the second more difficult part of the question.

WRONG: Why was the most recent Honduran government overthrown and explain what the government could do to prevent future overthrows?

RIGHT: (Ask two separate questions.)

WRONG: When, how, and why did World War II begin for the United States?

RIGHT: (Ask three separate questions, or at least two.)

PIT #4: Yes/No Questions

When you fall into this pit, you ask questions that must be answered by either yes or no, even though you don't want such an answer. Let's say you want to know who the Prime Minister of Canada is and you ask, "Mary, can you tell me who the Prime Minister of Canada is?" Mary answers, "Yes." Then there is an awkward moment. You know that her answer is technically correct and you also know that she really knew what you wanted and is probably toying with you a bit. What do you do?

You realize you phrased the question improperly (you blush a little) and then you ask the obvious next question, "Who is he?"

That sort of faulty phrasing goes on often in classrooms, so often that students are used to it. They know what their teachers really want in the first place and they usually give the appropriate answers. Because students do the expected, teachers never catch on to the fact that their questioning is a bit sloppy. Sloppy questioning leads to sloppy or inprecise thinking and it breeds problems. It also wastes precious instruction time. As teachers, we work against ourselves when we allow this to go on.

Unless you specifically want a yes or no answer, don't fall into PIT #4. In our examples below we'll assume you do not want yes/no answers. That assumption sets up our WRONG and RIGHT format.

WRONG: Could we say oil is the main export of Kuwait?
RIGHT: What is the main export of Kuwait?

WRONG: Is the hexagon larger than the octagon pictured on page 12?
RIGHT: Which one of these two geometric shapes on page 12 is larger? (Here the teacher probably wants the students to have practice in saying words like *hexagon* and *octagon*. They won't say them if they are asked yes/no questions.)

WRONG: Can you place the following four words in a sequence that will form a sentence: <u>soil, the, till, farmers?</u>
RIGHT: Form a sentence by placing the following four words in a sequence: <u>soil, the, till, farmers.</u>

Phrasing questions by using auxilliary verbs as the first word is what springs the trap that lets you drop into PIT #4. The typical auxiliary verbs are: Has . . . ? Is . . . ? Was . . . ? Can . . . ? Shall . . . ? Will . . . ? Would . . . ? Does . . . ? Have . . . ? Are . . . ? Were . . . ? Could . . . ? Should . . . ? Do . . . ? Begin a question with any of those words and you're headed for the pit. Think of two or three questions and then begin them with one of the auxiliary verbs. Get the feel for what happens.

WRONG: Should heads of state have military backgrounds?
RIGHT: (You do this one.)

WRONG: Is teaching the best profession there is?
RIGHT: (Do this one, too. Let's hope your *answer* to the question is also right.)

PIT #5: The Unquestion

Do you remember being asked the following questions or, worse yet, do you remember asking these questions?

How come?
Why's that?
And?
So?

Those phrases and one-word utterances really do not qualify as bonafide questions. At best they are "unquestions." Sometimes they get results but not everyone in the class is necessarily following the flow of questioning that led up to the "How come?" It is far, far better to take the time to construct a full question. In doing so the effective teacher makes sure she is beaming her questions to every student ensuring that everyone has the benefit of a fully stated and understood question.

No examples or warnings need to be posted at the mouth of this pit. It's enough to say, "Avoid the unquestions like the plague! Got it?" (Sorry about the question just asked. We all fall into #5 every now and then.)

Summary

This chapter was a very modest account of some of the basic information teachers need to know about questioning. Seven questions exercised your mind early in the chapter. The exercise led you to the realization that questions are critical to effective teaching. The Three-Quarters Rule was established as something with which teachers must cope.

Then came a six-way classification system which enables the teacher to hone in on desired levels of thinking. Using questions from varied levels of thought is more likely to educe varied levels of thinking from students, something that is vital to students' education.

Finally, you considered some of the WRONGS and RIGHTS of phrasing questions. Five rather common phrasing errors made by teachers were identified as labels for five pits. Unquestionably, all five are THE PITS!

PART III

Ending the Lesson

Chapter 11

CLOSURE#

Beginning the ending chapter seems paradoxical, yet all things must have a beginning and an ending. And whatever has a beginning and an ending must also have a middle. So it is with teaching.

A beginning, a middle, and an ending . . . so obvious, yet so elusive. As a result, teachers often do not develop their lessons with that idea firmly in mind. They typically begin the lesson without much forethought. They start somewhere, usually where they left off yesterday or where the next logical point of the subject matter dictates.

Beginning a lesson without a plan, as well as a strategy for attracting and intensifying the interest of the student, is an exercise in mechanical and often ineffective teaching. Effective teaching is done by a teacher who is mindful of the need for planned OBJECTIVES and SET INDUCTION that reach out and attract students to the learning process. Knowing the objectives and getting a proper inducement for achieving those objectives, students will be MOTIVATED to learn. That's "The Right Stuff" for beginning a lesson! Effective beginnings are always made with the student in mind.

Effective "middles" are also taught with the student in mind. The degree of effectiveness is determined by the way the teacher verbally and non-verbally REINFORCES student activity; the way a teacher ENCOURAGES students in their efforts to learn; the way

#"Closure" as a teaching skill was originally developed for use in a teacher training setting by W. D. Johnson for the School of Education, Stanford University, Stanford, California.

a teacher shows ENTHUSIASM for the subject as well as the act of teaching; the way a teacher brings a VARIETY OF STIMULI to the student; and the way a teacher poses thought-provoking QUESTIONS. Those things are all parts of effective middles.

Effective endings? That's where we are now in our analysis of effective teaching. This chapter is devoted to identifying what teachers must do to "close" their lessons effectively. At the same time we are analyzing CLOSURE we will be implementing closure on this book. In that sense the book models the process while it identifies the content of effective teaching.

Chapter Objectives

Understand the need for closure as an effective teaching skill.

Understand the components of closure.

Make a commitment to use closure in teaching.

Perhaps you remember from your initial study of psychology the Gestalt principle of closure. You learned a definition something like, closure is the principle that describes an individual's striving to see a broken or irregular figure as smooth and complete. The definition was learned in the context of visual perception, and you were shown examples like the two below:

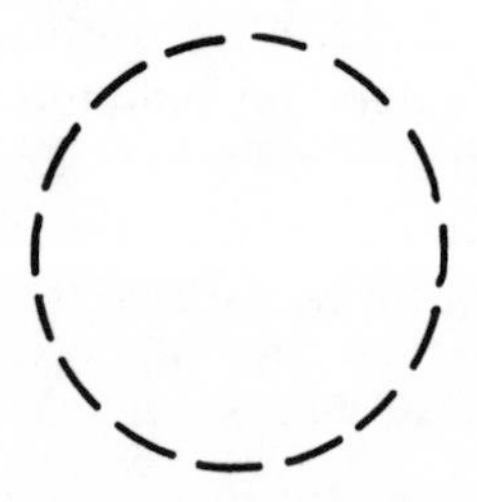

After studying the broken and irregular figures you were asked what you saw and you answered, "Circle and person-on-a-horse." Then you were led to explain why it is you saw those things when in reality they were not true drawings of a circle and a person-on-a-horse. Your answer and the professor's ensuing explanation led you to understand more fully what closure means and why it is that humans strive for seeing things smoothed out and complete or *closed*.

We can build on those early understandings of the Gestalt principle of closure and learn what the term has come to mean in the skill of teaching.

Before we get into our precise definition of closure we need to understand two things. One is that students will try to make sense out of what teachers do even when their teaching hasn't been clear. That's the nature of the learner—to strive to see "smoothness and completeness." It is one way humans learn. Such striving is already inside the learner. All the teacher need do is capitalize on it by using it to maximum advantage. Doing so will increase the effectiveness of the teaching.

Another thing about closure is that it is more often than not overlooked in the act of teaching. Closure is the most often ignored teaching skill of the nine skills analyzed in this book!

In 21 years of formally observing other teachers at work, I have personally seen lessons—many lessons—too many lessons begin and end without a hint of planned closure. Numerous principals and supervisors of teachers have told me the same thing. And closure is one of THE most important skills!

What is closure in teaching?

Why is it so important?

Definition: Closure is teacher or student activity that brings the main points of a lesson into focus so that they may be perceived as an organized whole.

Critera of Effective Closure:

—reviewing, summarizing, organizing
—student demonstrated achievement
—connecting past, present, future

Reviewing/Summarizing/Organizing

In this criterion of closure we note that effective teaching must include activity which draws attention to the completion of the lesson or a significant part of the lesson. Students may or may not have a sense of the organization or completion of the lesson or its significant points. Regardless of what the students sense about closure, it is the teacher's responsibility to make sure all students have the "sense."

Teacher activity that includes the review, summary or organization of material just covered helps students develop a sense of completeness and mastery. The teacher may do the reviewing, summarizing, or organizing but the impact on students is significantly increased whenever *they* do the work. At first it will likely be a labor intensive effort to get students to do the work. It will be like the proverbial "pulling of teeth." However, with consistent use of this practice and with the necessary encouragement to students to do the work, it will get better. Students will know you expect this procedure of them and they will live up to your expectations. They will be better for it. They will learn more!

Many teachers, probably most teachers, neglect this critical skill. They allow the end-of-the-period bell to "close" their lesson or they allow a glance at their watch to prompt them to exclaim, "Oh, wow! There's only a minute left!" or they ask that all too typical "Any questions, class?" There being none, teachers assume student mastery of the material and move on to the next lesson segment.

The activities of summarizing, reviewing, and organizing are of course different. All three need not be used for drawing student attention to the completion of the lesson or of significant parts of the lesson. Whichever activity or combination of activities will provide students with the sense of closure ought to be used.

Before we look at an example of the first criterion of closure, let's look at the second criterion. After understanding the second, we'll study an example that will illustrate the first two criteria simultaneously.

Student Demonstrated Achievement

This is one criterion most teachers meet, yet when they are asked why they are doing it, they are hard pressed for a sound answer. If we ask, "Why do you have students recite or demonstrate what they have learned?" we get something like, "How else will I know if they're getting what I'm teaching?"

Even though our question was answered with another question, we realize the teacher has the right idea. However, teachers must have more than "the right idea." They must know why a particular procedure is used. Furthermore, they must have the skill to implement the procedure that brings about the desired result—in this case, mastering the material taught.

By having students demonstrate their achievement, in the teaching-learning setting, several things are being done. First and foremost, the students who are demonstrating their achievement are learning more. Their demonstration provides them an opportunity to reproduce what was taught. In doing so, the learning is stamped in, so to speak. Being on the receiving end of something being taught is one thing. Being asked to be on the demonstrating end is another thing. We all know that when we act on something we have internalized, we know it better.

Ask experienced teachers what the effect of teaching a particular subject is on their understanding of their subject. They will tell you in no uncertain terms, "I learned my subject better in getting ready to teach it than I did when I learned it in the first place." Teaching one's subject to another is doing something with it. It is a way of demonstrating achievement. It is the teacher's closure.

Students' demonstrating their achievement of material learned in a lesson is analogous to the teacher's teaching of material to students. The point is that whenever students do something with the material just covered, they learn it better. It is the students' closure.

Another thing accomplished by student demonstrated achievement is personal recognition. The student who is provided the opportunity to demonstrate, to his peers and his teacher, is most certainly getting a motivational shot in the arm. In Chapters 2 and 3 we saw the intrinsic value of recognition in the motivation of the student. Motivation and learning are inextricably bound together..

A third accomplishment of this teaching procedure of closure is the opportunity for teachers to evaluate their teaching effectiveness. Should teachers find that students can readily demonstrate what was taught, they can move on to the next segment or the next lesson with confidence. Should teachers find that students cannot demonstrate what was taught, they can make the necessary adjustments in their teaching so that student learning will occur. Too often teachers *assume* that their students have learned, when they have not.

There are all sorts of examples you can think of for getting students to demonstrate their achievement of the lesson points. Some of the more obvious ones are:

(1) Asking the students to recite what they have learned so far.

(2) Directing one student, then another and another, etc. to summarize what was covered. Multiple participants will more likely insure adequate closure.
(3) Giving a mid-lesson written quiz.
(4) Sending students to the chalkboard to write what they have learned, via solving problems, spelling words, drawing pictures or designs, etc.
(5) Making a homework assignment or a seatwork assignment that the students begin during the class period. The teacher circulates among the students assessing their work.
(6) Sending the student(s) to a lab station to do an experiment or to make something that will illustrate their grasp of the material taught.
(7) (Not so obvious) Asking all students to close their eyes and respond to the following questions by raising their right hands for TRUE and their left hands for FALSE for statements the teacher makes about the material. Closed eyes will prevent other students from influencing one another.

As a further example, let's ask you to effect closure on what you have learned from reading this book.

Please name the nine skills covered in the eleven chapters of this book.

In your answer to that question, you were employing the first two elements of closure, namely REVIEWING/SUMMARIZING/ ORGANIZING and DEMONSTRATING STUDENT ACHIEVEMENT.

Now, here's an even more difficult assignment for you. See if you can summarize and organize an outline of those nine skills, complete with their sub-elements or critera. To help in this rather difficult assignment, an outline will be provided. You need only write in words that identify the criteria.

OUTLINE OF EFFECTIVE TEACHING SKILLS

I. Motivation

t_c ______________________________

C ______________________________

I ______________________________

R ______________________________

II. The Objective

Verbs which typify the cognitive domain

K ______________________________
C ______________________________
A ______________________________
A ______________________________
S ______________________________
E ______________________________

III. Set Induction

I ______________________________
C ______________________________
U ______________________________
M ______________________________

IV. Reinforcement Cueing

V ______________________________
N __________ V __________________

V. Encouragement

B ______________________________
R ______________________________
R __________ C __________ to S __________
O __________ A __________
U __________ A __________
E __________ S __________ A W __________
NonA ____________________ of M __________
C ______________________________

VI. Enthusiasm

Enthusiasm for one's S __________________
Enthusiasm for one's T __________________

VII. Stimulus Variation

M ______________________________
G ______________________________
F ______________________________
S __________ S __________ C __________
P __________

(continued)

OUTLINE OF EFFECTIVE TEACHING SKILLS (continued)

VIII. Questioning

Levels K ______
C ______
A ______
A ______
S ______
E ______

Pitfalls
#1 S ______ — F ______
#2 A ______
#3 The C ______ Q ______
#4 Y ______ / N ______ Q ______
#5 The U ______

IX. Closure

R ______ /S ______ /O ______
S ______ D ______ A ______

Not yet developed:
C ______ P ______ P ______ and F ______

In completing that outline, you and I achieved an important closure. Perhaps you felt a sense of completeness, a sense of mastery over it all. At least you "got a better handle" on the total picture. That sense, that handle, is the effect of closure.

If you're also sensing a feeling of being overwhelmed by all nine skills, with all their components, don't let it bother you. It's like anything else that is important and comprehensive. Once you get into it and use the skills, making some errors and achieving some good results, it will all become clearer and better. The Chinese have a saying that seems appropriate here. "The 1,000 mile journey begins with the first step."

On to the third criterion of closure. . . .

Connecting Past, Present, and Future

Reflecting on the above topic, you can quickly get the idea that helping students see how today's lesson ties in with the past and the future gives the idea of "an organized whole." You will recall that in our definition of closure we said that the student had to perceive an organized whole in order for closure to occur. Perhaps you have picked up on the fact that in various parts of this book *you* were asked to think of your past experiences. You were also asked to imagine how you would do certain things. These references were to your past and your future. The attempt was made to have the point under study *connect* to your experiences so *you* would feel closure. Feeling a sense of closure ought to convince you of the need for closure in the classroom.

Did you notice how some of the chapters began as a set induction for the material about to be developed? Included in the chapter sets was a tie-in to your past and future experiences. (This book, so far as possible, exemplifies the tenets of effective teaching.) When the chapter sets did tie in with your experiences, closure was accomplished by way of its third criterion, connection of past, present, and future. Closure and set induction are complementary in this regard. We'll look at this idea at the end of the chapter.

Let's illustrate an actual case of a teacher effecting closure with her students using this third criterion. The example cited comes from an actual teaching observation of a fifth grade teacher who was teaching a unit on first aid.

> *Teacher*—"Boys and girls, if the forearm is broken, great care must be used to keep the broken area from being bumped or moved unnecessarily. You will remember last week we were studying bones of the body. What is this particular forearm bone called?"
>
> *Several Students*—"The ulna!"
>
> *Teacher*—"Thank you! Well, when we go to the Reading Hospital next week we will visit the emergency care unit. There we will learn how we can help someone who has a break in that area of the arm."

Here we see the connection of the three time elements.

Effective teachers even go so far as to remind students how today's lesson ties in with what they learned two or three years ago and how it will help them, clearly and specifically, at some time in the future. Students need to perceive the flow and organization of things learned over time and of course need to perceive the relevance of it all. Closure helps fill this need.

Closure is needed, not only in daily lessons, but also in other units of teaching. These units include the following:

(1) units of study that extend over days or weeks
(2) films, filmstrips, audio or video tapes, recordings
(3) drill sessions, question and answer sessions, discussions

(4) field trip follow-up
(5) guest speaker follow-up
(6) laboratory experience
(7) homework or seatwork assignments

Stated earlier was the idea that closure and set induction are complementary. A quote taken from T. S. Eliot will make this point very well.

> "What we call the beginning is often the end and to make an end is to make a beginning. The end is where we start from."

Frequently teachers will begin a new lesson with a summary of the previous day's lesson. When they do, they are still working to achieve closure of material previously taught. The skill being used is closure, even though the teaching is being done at the beginning of the lesson. However, if the closure generates a desire on the part of the students to a point where they WANT to get into the upcoming lesson, the closure serves as a set induction. In cases like these the teacher is effecting closure and set induction simultaneously!

Simultaneous use of closure and set induction is a desirable practice. Closure generates a sense of accomplishment in students. That sense of accomplishment further generates an intrinsic motivation to learn and of course motivated students WANT to learn. Hence, we see the complementarity of the two skills.

So now we come to the ending of this chapter on closure. It's time to bring further closure to closure.

By way of review, summary and organization, we covered the following points that tied in with the chapter objectives:

(1) Closure is a vital teaching skill. If students are to derive maximum benefit from the material taught they need to have a feeling of completion and organization for the material. Further, they need to have a sense of mastery over it.

(2) Closure has three components that serve as criteria for determining whether the teacher is working to effect closure to the lesson, unit, discussion, audio-visual presentation, etc. The three components are:
   a. reviewing, summarizing, organizing
   b. student demonstrated achievement
   c. connecting past, present, and future

(3) As for the third objective, namely your making a commitment to using closure in your teaching, you engaged in some activities that can lead to the commitment. You answered some questions, you fulfilled some assignments that gave you a sense of achievement, a sense of mastery, over closure. Feeling that sense of mastery, you will surely want to try out closure in the live teaching setting.

When you answered the question, "What are the nine skills covered in the first eleven chapters?" You were demonstrating your achievement. When you worked to complete the outline of the nine skills and their sub-components you were once again demonstrating your achievement. In doing these things, closure was being made.

We began this chapter with a review of the prior ten chapters. The review tied the current chapter to *past* chapters. You were encouraged to use the skill in your *future* teaching endeavors. Past, present and future experiences were brought to bear in this chapter. By doing this, the third criterion of closure was implemented.

There we have it . . . closure on closure.

Chapter 12

EPILOGUE

You really *have* chosen the most important profession in the world! Teaching *truly* is the queen mother of professions. After reading the preceding eleven chapters, you must have an enhanced image of the vital work of the teacher and your specific role in it.

Skills required of effective teaching have been identified and explained in detail in this book. Since you need a knowledge and an understanding of those skills, as well as their component parts, you no doubt feel a challenge as you contemplate how you will go about developing the skills involved. Teaching, like any other profession, requires its practitioners to master certain requisite skills. Mastering the skills of the profession and using them effectively is the embodiment of true professionalism. Not everyone is capable of mastering them.

There is a notion in the public domain that anyone can teach. Just take someone who knows his subject matter well and can relate well to people and there you have it, a person who can teach. That kind of thinking is dangerous because it does contain two of the elements for the making of an effective teacher: adequate knowledge of the subject to be taught, and an understanding of the nature of the learner.

The missing element for effective teaching is the body of skills involved with the planning, preparation and presentation of the lesson. Some refer to this as methodology, others call it pedagogy, still others call it techniques, strategies, and practices. For our purposes, the third element is EFFECTIVE TEACHING SKILLS.

There is a body of skills that one must master if he or she is going to bridge the gap between the learner and the subject to be taught. Not everyone is capable of mastering those skills. Some people believe one has only to stand in front of a group of students and tell them what they need to know and they'll learn.

A teacher cannot just talk about the subject to the learner and expect him to learn it. Teaching at its best is not talking about or telling. It is far, far more. One of the posters circulating among educators puts it rather well.

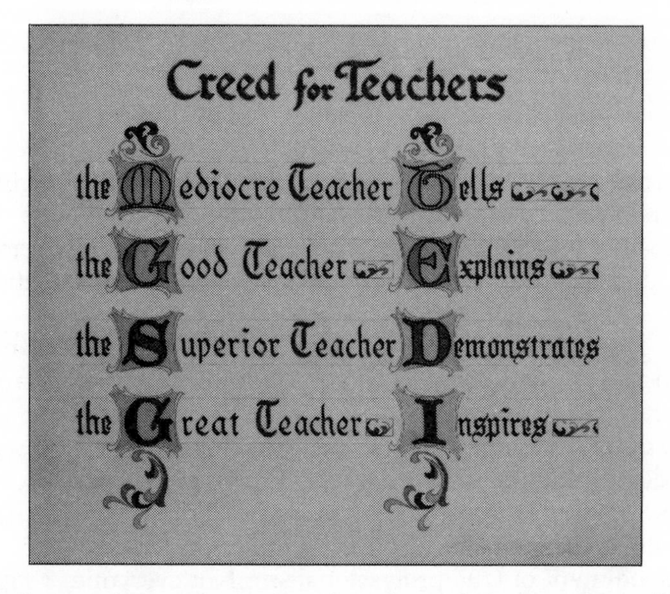

Good, superior, and great teachers have demonstrable skills that enable them to be good, superior and great. The identification of those teaching skills has been a longstanding quest of the committed educator. The nine skills identified in Chapters 2–11 are a part of that body of effective teaching skills. Certainly the final word on effective teaching has not been printed in this or any other book, but the nine identified in *Effective Teaching* are valid, research-verified skills. When used, they consistently heighten the probability that the student will indeed learn.

Consistent effectiveness in teaching is achieved only by the committed professional teacher. The professional teacher must be committed to learn and master a body of teaching skills that when practiced will bring student and subject matter together to a high level of productivity.

In education, productivity is student learning.

Bibliography

Bloom, Benjamin S., ed., *Taxonomy of Educational Objectives, The Classification of Educational Goals, Handbook I: Cognitive Domain,* New York: David McKay Company, Inc. (1956).

Bruner, Jerome, *Toward A Theory of Instruction,* New York: W. W. Norton, Inc., pp. 113–138 (1966).

DuBelle, Stanley T., Jr. and Carol M. Hoffman, *Misbehavin'* and *Misbehavin' II,* Technomic Publishing Co., Inc., PA (1984).

Fritz, Jean, *George Washington's Breakfast,* Coward-McCann, Inc., New York

Gillett, Maxwell H., "Effects of Teacher Enthusiasm on At-Task Behavior of Students in Elementary Classes," Doctoral dissertation, Eugene, OR, University of Oregon (1980).

Harrow, Anita J., *A Taxonomy of the Psychomotor Domain,* David McKay Company, Inc., New York (1972).

Hunkins, Francis P., *Involving Students In Questioning,* Allyn and Bacon, Inc., Boston (1977).

Hunkins, Francis P., *Questioning Strategies and Techniques,* Allyn and Bacon, Inc., Boston (1972).

Hyman, Ronald T., *Strategic Questioning,* Englewood Cliffs, New Jersey: Prentice-Hall (1979).

Krathwohl, David R., et al., *Taxonomy of Educational Objectives, The Classification of Educational Goals, Handbook II: Affective Domain,* New York: David McKay Company, Inc. (1964).

Phi Delta Kappa's Center on Evaluation, Development, and Research, Volume 3, No. 4, Indiana, (June, 1981).

Rosenshine, Barak and Norma Furst, "Research on Teacher Performance Criteria," in *Research in Teacher Education: A Symposium,* B. O. Smith, ed. Englewood Cliffs, N.J.: Prentice Hall (1971).

Sanders, Norris, *Classroom Questions: What Kinds?* New York: Harper and Row (1966).

ABOUT THE AUTHOR

Stanley T. Dubelle—In a 31-year period of time, Dr. Dubelle has been a classroom teacher, athletic coach, school administrator, college professor, and superintendent of schools. He graduated with an A.B. in History from Franklin and Marshall College, an M.A. in Classroom Teaching from West Virginia University, and a Ph.D. in Secondary Education from The Pennsylvania State University.

He has served as a consultant-speaker for 50 different educational institutions. His work includes topics on establishing a positive learning atmosphere, teaching skills, student motivation, as well as developing student self-discipline in the schools.

He has authored approximately 100 articles for newspapers, magazines, and journals. He is the co-author of the books *Misbehavin'* and *Misbehavin' II.*

In 1980, Dr. Dubelle was chosen Outstanding Educator by the Penn State Alumni Association. He is married and the father of three children.